I0624252

Future-Proof

How to Adopt and Master Artificial
Intelligence (A.I.) to Secure Your Job
and Career

Philip Blackett

Preface

"The future depends on what we do in the present." -
Mahatma Gandhi

In a rapidly evolving world where artificial intelligence (A.I.) is reshaping industries and redefining the future of work, it's easy to feel overwhelmed or uncertain about the future of your career. *Future-Proof: How to Adopt and Master Artificial Intelligence (A.I.) to Secure Your Job and Career* is a guide crafted to empower working professionals, both white-collar and blue-collar, to harness the power of artificial intelligence and machine learning to further their careers and protect them from job insecurity.

This book delves into the realm of AI, offering insights, strategies, and practical advice to help you not only adapt to technological changes but also to thrive and become indispensable linchpins in your team, department, and company. The purpose of this book is not just to equip you with knowledge, but to instill a sense of confidence and purpose as you navigate the ever-evolving landscape of the workforce powered by artificial intelligence.

As the author, I've been driven to write this book by the real and often poignant experiences of individuals like yourself who face the uncertainty of job security in a world increasingly influenced by technology. Take, for instance, the story of Jessica, a dedicated customer service representative who found herself constantly worrying about the potential automation of her role. Her journey and the recent stories in January 2024 of widespread corporate layoffs due, at least in part, to artificial intelligence have inspired the creation of this book – a beacon of hope and guidance in uncertain times. I want to help the Jessicas of the world prepare for what is already happening around them with the proper mindset and tools to succeed in this new world.

They say wisdom comes from many sources, and this book has also been greatly influenced by the invaluable support and contributions of experts in the fields of AI, technology, and business. Their insights have been invaluable in shaping this comprehensive guide, ensuring that it not only inspires but also imparts practical wisdom.

To you, I extend my heartfelt gratitude for considering this book as part of your journey toward professional empowerment and skills development in a workplace powered by artificial intelligence and machine learning. Your time and attention are precious, and I am deeply honored that you have chosen to invest them in this work and in your career progression.

The wisdom within these pages is intended for anyone who seeks to embrace the future with courage and tenacity. Regardless of your background or current level of understanding of AI, this book is designed to be accessible and comprehensive, providing a wealth of knowledge and strategies to propel you forward.

As you delve into the chapters that follow, I invite you to embrace the lessons, insights, and strategies they hold. By the end of this journey, it is my hope that you will emerge with a newfound understanding of AI, equipped with the tools to enhance your skills, elevate your productivity, and fortify yourself against the uncertainties of the ever-changing professional landscape.

Thank you for your investment in this book. Let its wisdom be the guiding light on your path to professional fulfillment and job security.

Contents

Chapter One

Embracing AI as an Opportunity, Not a Threat

I n the soft light of a San Francisco tech hub, Ava sat encircled by a garden of screens, each one a window into a realm of artificial intelligence and machine learning. She toggled between webinars and code, caught in a limbo between wonder and apprehension. The thrill of progress filled the room like incense, but beneath it lingered a persistent uncertainty. Could her skills weather the storm of change that AI promises to unleash?

She paused, her fingers hovering over the keyboard as she considered the algorithms she was weaving, ones that could one day supersede her own current abilities. The sun quietly surrendered to dusk outside her window, reflecting her own resignation to the march of time. Yet, Ava's faith reminded her that every innovation under heaven had its season. As AI augmented human capability, so too could it amplify the divine spark within her, catalyzing her growth rather than supplanting her essence.

With a sip of her cooling coffee, she contemplated how Elijah, moved by the voice of stillness, found his strength not in the overtaking wind, the earthquake, nor in the fire, but in the gentle whisper that followed. This AI, her modern-day whisper, was an invitation to partnership and not the ominous threat some of her peers saw it to be. She remembered the parable where servants were given talents; some buried them out of fear, but the wise multiplied them. Ava intended to invest her talents wisely.

Across the office, her colleague lamented over his inability and unwillingness to adapt to new technology, his words a heavy stone that sank in the room's atmosphere. Ava's resolve, however, was buoyed by entrepreneurial spirit and a belief that the wise adapt like the reeds to the wind, finding strength in flexibility. Her mind was a garden; AI, the new seed she was learning to nurture alongside the knowledge she had already cultivated. Her own hands, guided by conviction, seemed almost to become extensions of the GPTs she programmed, her human creativity fusing with digital precision in a symphony of potential opportunities.

The hum of servers became a chorus, bolstering her conviction that the bridge between human and artificial intelligence was one to be crossed with boldness. In the teachings of faith, she sought parallels, wisdom that held true even as the world shifted beneath her feet. The marketplace, after all, was always a mosaic of change, and those who wove themselves into its fabric with adaptability and foresight were those who survived such change and later penned history.

As Ava turned from her work, her gaze settling on the orange hue of street lights flickering to life, a question lingered, mirrored in the depths of the evening's onset. Could there be a more pressing task than to weave one's intellect with the threads of technological progress, finding in the partnership with artificial intelligence (A.I.) not a competitor, but a collaborator in the dance of growth within Silicon Valley?

The Dawn of Collaborative Intelligence

At the heart of every technological revolution is the potential for human advancement. This was the case with the wheel, electricity, the telephone, the automobile, the airplane, the railroad, the Internet and social media. As we stand on the precipice of the artificial intelligence era, **we are faced with a choice: view AI as a harbinger of job insecurity or embrace it as a catalyst for professional enrichment**. The wisdom inherent in spiritual teachings encourages us to seek understanding and growth in all facets of life, and in AI, we find a powerful tool for both. AI can be an ally, amplifying our professional capabilities and leading to a higher level of service and innovation in our work and in our careers.

No discussion about the future of careers can be had without acknowledging the pervasive anxiety surrounding AI's role in the workplace. But let's consider AI's potential from a perspective of empowerment and abundance, rather than from fear and scarcity. AI's rapid advancement isn't a signal to retreat but a call to evolve, urging us to **redefine the partnership between humans and machines**. By enhancing our unique human skills, we can ensure our professional relevance and demonstrate our unforgettable value to employers.

Diving into this rich subject, we find three keys to unlocking the benevolent power of AI in our professional lives. First, **we must understand how AI can augment our skills**; not replace, but enhance the work we do, allowing us to operate at new heights of efficiency and creativity. Secondly, **collaboration with AI should be viewed as a strategic partnership**, where our human intellect and emotional intelligence synergize with AI's data processing capabilities to yield unprecedented outcomes. And lastly, **we must be open to growth and adaptation**, leveraging AI as a means of professional and personal development.

As we confront the challenge of securing our careers in a rapidly changing environment, let us be reminded that wisdom lies in understanding the times we are now in and adapting to them. AI is no exception; it is another step in our collective journey, presenting both challenges and answers. This chapter will serve as the foundation, illustrating the ways in which AI can become a formidable ally in the pursuit of your career longevity and professional success.

Drawing upon theological wisdom, economic principles, and political frameworks, our discourse will weave a rich tapestry of knowledge, essential for navigating the AI landscape with confidence and grace. We will also uncover the essence of future-proofing your career in an AI-driven world.

The subsequent chapters will build upon this foundation, exploring how to master AI tools, prepare for the jobs of the future, and cultivate a mindset of continuous learning and adaptation. Your journey through this book will equip you with the knowledge to stand unwavering in the face of technological change, positioning yourself as an indispensable part of the workforce.

Let us embark on this exploration with humility and determination, recognizing that in our hands lies the power to shape our destiny in harmony with the tools we create. AI is not the end of the working professional—it's the beginning of a new chapter in human ingenuity and collaboration with the latest in technology.

Artificial Intelligence (AI) and machine learning are widely regarded as disruptive technologies that have sparked concerns about job displacement among working professionals. Many fear that their current skills will become obsolete, rendering them redundant in the job market. However, the truth is quite the opposite. **Rather than posing a threat, AI presents an opportunity for professionals to augment their skills and enhance their performance in the workplace**. By leveraging the capabilities of AI, individuals can elevate their value, secure their place in the job market, and advance their careers.

AI functions as a collaborator, not a replacement. It equips professionals with tools to streamline processes, analyze data with greater accuracy, and make more informed decisions. Think of it as a partnership that empowers individuals to exceed their current limitations and soar to new heights in their professional endeavors. Embracing AI is not about ceding control or bowing to a new overlord; it's about leveraging technological advancements to enhance one's skills, performance, and value in the workplace.

When professionals embrace AI, they become more resilient and adaptable in an ever-evolving job landscape. Rather than fearing obsolescence, they can proactively position themselves as indispensable assets to their employers. This proactive approach will not only solidify their place in their current roles, but also open doors for career advancement and increased opportunities within their organizations.

Moreover, AI has the potential to optimize workflow efficiency, expedite tasks, and generate valuable insights that traditional methods may overlook. By harnessing the power of AI, professionals can become more agile, innovative, and effective in their roles and responsi-

bilities, demonstrating an unparalleled ability to drive higher-level results and add significantly more measurable value to their companies.

**In today's uber-competitive job market, the ability to har-
monize with AI and extract maximum benefit from its capabilities is a potent differentiator**. It sets professionals apart as
forward-thinkers who harness the potential of technology to enhance
their own abilities and contribution to their teams. By embracing
these technologies, individuals can steer their careers toward greater
success, job security, and fulfillment in the workplace.

Embracing AI is not about relinquishing control, but about gaining mastery over new tools and technologies. It's an opportunity to
unlock new levels of potential and elevate professional performance.
Let's explore how collaboration with AI can further enhance professional value and relevance in the job market.

AI and machine learning are not threats to the job market; rather,
they offer a prime opportunity for professionals to elevate their value
and relevance. By collaborating with AI, individuals can harness its
capabilities to expand their skills and knowledge, ultimately increasing
their worth in the job market. Embracing this integration can lead to
improved performance, expanded career opportunities, and height-
ened job security.

Leveraging AI to Enhance Professional Value

Collaborating with AI technology enables professionals to amplify
their capabilities and bring more value to their roles. By embracing
AI as a tool for personal and professional growth, individuals can take
advantage of its potential to streamline processes, generate data-driven
insights, and automate routine tasks. This collaboration empowers
professionals to focus on higher-level strategic initiatives and creative problem-solving, contributing to their overall effectiveness in the
workplace.

Augmenting Skills and Knowledge

The integration of AI into professional practices can lead to an augmentation of skills and knowledge that positions individuals as valuable assets in the job market. As professionals work alongside AI, they have the opportunity to refine their expertise, expand their understanding of data analysis and interpretation, and gain proficiency in leveraging cutting-edge technologies. This continuous learning and collaboration with AI bolsters their adaptability and relevance, vital qualities that employers seek in a rapidly evolving professional landscape.

Increased Competitive Advantage

Embracing AI as a collaborative partner offers professionals a competitive advantage in the job market. By harnessing the insights and efficiencies that AI provides, individuals can distinguish themselves as forward-thinking, innovative professionals who are adaptable to technological advancements. This positions them as desirable candidates for career opportunities and enhances their overall marketability, ensuring a stronger foothold in the competitive job market.

Paving the Way for Career Growth

The integration of AI into professional practices can serve as a catalyst for career growth and advancement. Professionals who actively collaborate with AI position themselves as thought leaders in their respective fields, contributing to the development of new strategies, innovative solutions, and enhanced productivity. This involvement paves the way for professional growth, as individuals become instrumental in shaping the future of their industries and organizations through their collaboration with AI.

Setting the Stage for Job Security

Embracing AI as a collaborative partner not only elevates professional value but also sets the stage for enhanced job security. By actively engaging with AI and leveraging its capabilities to drive results, professionals become indispensable contributors to their organizations. This essential role as a value-adding collaborator fortifies their position within the company, increasing their job security and reducing the risk of displacement by AI-related automation.

In summary, the collaboration with AI offers professionals the opportunity to enhance their skills and knowledge, gain a competitive advantage in the job market, pave the way for career growth, and fortify their job security. By recognizing AI as a complementary tool for their personal and professional development, individuals can position themselves as indispensable assets well-equipped to thrive in an AI-enabled workplace.

In a world where artificial intelligence is becoming increasingly pervasive, harnessing its capabilities can be an extraordinary step toward personal and professional growth. For those willing to explore the potential of AI, the key is to view this technology not as a competitor, but as a collaborator—an extension of one's own talents and abilities. AI, when engaged effectively, has the power to amplify your expertise, sharpen your insights, and expand your professional reach. This begins with approaching AI with an open mind and a willingness to learn the language of this new digital helper.

Reframe Your View on AI

Embracing AI starts with a shift in perspective. Think of it as a partnership where you and artificial intelligence bring unique strengths to the table. **Your creativity, critical thinking, intuition, and emotional intelligence combined with AI's speed, accuracy, and vast data processing capabilities create a synergy that can tackle complex problems with efficiency and innovation**. Remember the words of the wise King Solomon, "Two are better than one; because they have a good reward for their labor." (Ecclesiastes 4:9). By pairing human ingenuity with artificial intelligence, you can discover solutions that neither could achieve alone.

Cultivate AI Literacy

To work effectively with AI, commit to becoming AI-literate. This involves staying updated on the latest advancements and tools, understanding the basic concepts of machine learning, and recognizing the practical applications of AI in your field. Seek knowledge from a variety of sources—online courses, books, conferences, workshops, webinars—and network with professionals who are also embracing AI. By actively seeking understanding, you prepare yourself to employ AI as a powerful aid in your undertakings.

Integrate AI into Daily Workflows

Begin by integrating AI gradually into your daily workflows. Start small—employ productivity tools enhanced by AI, or streamline tasks with automation software. As you witness firsthand how AI can alleviate the burden of repetitive tasks, your confidence will grow. Allow AI to handle data analysis, for instance, while you focus on strategic decision-making. The result? You contribute at a higher level, bolstering both your effectiveness and your value in the workforce.

Embrace Continuous Learning

In the dynamic realm of AI, the only constant is change—a trait commonly shared with the world of business. **Continuous learning isn't**

just recommended; it's necessary to stay relevant. Your journey with AI should be marked by constant advancement and improvement.

Collaborate and Co-create with AI

In collaboration with AI, you can explore new horizons of creativity and innovation. Design with AI tools, develop new products, or enhance customer experiences. AI can serve as a creative partner, providing you insights drawn from data patterns impossible for the human brain to process alone. Let AI hone your professional skills, enabling you to deliver exceptional value in your endeavors.

Overcome Challenges with a Growth Mindset

Face AI-related challenges with a growth mindset. **Instead of shying away from unfamiliar technology, embrace the opportunity to learn and grow from it.** Tackle AI's complexities, don't be afraid of being uncomfortable in the process, and transform your initial challenges into stepping stones for advancement.

Redefine Your Unique Value Proposition

Finally, redefine and strengthen your unique value proposition in an AI-augmented workplace. Identify the skills and attributes that set you apart—your creativity, strategic thinking, leadership, empathy—and refine them. AI may analyze data at incredible speeds, but your unique human qualities bring meaning and context to that data. By accentuating these traits, you ensure that your professional profile remains indispensable, no matter how advanced AI becomes.

By embracing AI with a spirit of optimism and a commitment to lifelong learning, you prepare yourself not just to survive, but to thrive in an AI-enhanced future. **Remember, AI is a tool; you are the craftsperson.** Use it wisely, and watch as it amplifies your capabilities, opening doors to new possibilities and securing your place in the vanguard of your profession.

In embracing the potential of AI as an opportunity, not a threat, we align ourselves with a profound principle present in many spiritual teachings: the concept of growth through collaboration and adaptation. Just as spiritual texts emphasize the importance of working alongside others to foster personal and collective growth, so too can we tap into the power of AI to enhance our professional skills, increase our value in the job market, and promote personal and professional growth.

As we journey through the chapters ahead, we will delve deeper into the strategies and practical applications of AI that can catapult us into a future where we are not just secure in our careers, but thriving in them. Drawing inspiration from the wisdom of various disciplines, we will unearth the transformative potential of AI in the professional landscape. Through real-world examples, actionable advice, and profound insights, we will unlock the means to position ourselves for unparalleled success in a world of constant technological advancement.

With the tools, strategies, and insights presented here, you are on the brink of a powerful transformation. The future is not a distant, uncertain horizon; rather, it is a canvas upon which you will craft your success. Seize this moment and join me on this enlightening journey through the transformative landscape of AI mastery.

Chapter Two

AI's Relevance Across Diverse Fields

The sun's gentle farewell painted the city skyline with hues of gold and crimson, highlighting the silhouettes of silver-tipped towers. Amidst this tranquil backdrop, Elijah walked the bustling streets of New York, the city's heartbeat synchronizing with the thoughts racing through his mind. A seasoned businessman, he contemplated the transformative power of artificial intelligence—how it had begun to seep like rain into the soil of industries far and wide, far beyond his own field of financial services.

Elijah had always intertwined his faith with his business practices, believing that the guidance from sacred texts could align with cutting-edge technology. As he passed a small congregation leaving an evening service, the vibrant melodies of their songs reminded him of a verse that spoke of wisdom inscribed in the soul by the Creator, a wisdom now being mirrored by algorithms and data patterns. It was humbling and invigorating—where once he advised clients with financial insights drawn from a well of experience, he could now augment his advice with the depth of knowledge AI provided.

Turning a corner, Elijah's footsteps echoed against the stone as he entered his favorite café, a warm haven that had nurtured many of his future entrepreneurial plans. Over an aromatic cup of coffee, the steam rising like prayers, he mused over how AI could reshape content creation, marketing strategies, and customer interactions. The challenges his clients faced in sales and operations could be met with an intelligence that bolstered creativity and efficiency. It felt like a fresh wind blowing through the stained glass windows of a stately cathedral, awakening the air within.

His company had seen leaders hesitate, often wary of the unknown. Yet, as he shared personal anecdotes of AI's benefits, there was a shift. The new systems empowered them to thrive in an economy where big data and machine learning analytics influenced decision-making. It was vital for Elijah to convey the message that AI was not a force to be feared, but embraced, as it was an instrument poised to enrich human capability, not replace it, unless you're one that hesitates to adopt it.

As the day succumbed to the starlit embrace of the night, Elijah reviewed his notes for tomorrow's presentation, messages of inspiration blended with strategic planning. He poised his pen above the paper and inscribed a phrase that seemed both a revelation and a guiding principle: "Collaborate with the artificial, but nurture the human spirit." Would businesses grasp the full potential of artificial intelligence as a partner in their endeavors, and could they nurture its growth alongside the indomitable and compassionate human spirit?

Beyond Binary Codes: Unveiling AI's Universal Influence

Artificial intelligence (AI) has transcended the binaries of computer code to become an omnipresent force in our lives. The quest to master AI is not exclusive to the technologically adept; it beckons professionals of all disciplines to embrace innovation and thereby secure their place in the future of the workforce. The doctrine of value creation through AI holds the promise of transforming everyday occupations into strongholds of opportunity, granting those who harness it a form of vocational enhancement.

In a world where change is the only constant, AI stands as a lighthouse, guiding the multitude of industries through the fog of obsolescence. The breadth of AI is profound, touching the nuanced practices of content creation, the dynamic shifts of marketing strategies, the persuasive arts of salesmanship, the responsiveness of customer service, the analytics of finance, and the efficiency of operations. Recognizing the universal applications of AI serves as the first beacon towards modern-day professional enlightenment. The once siloed expertise in fields living in Silicon Valley now finds a shared destiny with the global digital frontier.

Discovering the possibilities for value creation through AI is tantamount to exploring new lands enriched with untapped resources. Entrepreneurs can make their return to relevance by leveraging AI to innovate and revitalize their businesses. The strategic usage of AI predicts not just incremental improvement but exponential growth.

Our capacity to identify opportunities to leverage AI is instrumental in prophesying a career resistant to the ravages of technological displacement. **Consider the parable of talents (Matthew 25:14-30); AI is the talent entrusted to today's professionals. Those who**

wisely invest it in their fields of expertise will reap manifold returns, while those who bury it in the ground stand to gain nothing. Content creators, for instance, can employ algorithms to personalize narratives, marketers can harness data analytics to predict consumer behavior, and customer service can utilize intelligent chatbots to enhance the user experience.

By infusing operations with AI, businesses not only streamline processes but also unearth innovative approaches to problem-solving. Finance professionals who introduce AI into their analysis can uncover insights of a magnitude previously hidden.

Employing an instructive voice, this chapter aims to shepherd professionals through the implementation of AI in varied terrains of enterprise. Like Ruth gathering barley in the fields of Boaz, individuals must actively glean the technological bounty that surrounds them, translating these grains of innovation into a harvest of career potency. Achieving mastery over AI tools can be a pilgrimage we set upon, undaunted by the language of prompts or the shroud of code, emboldened by the spirit of perseverance and the pursuit of wisdom.

In traversing this digital landscape, let us consider the marketer who, through predictive analytics, discerns patterns invisible to the naked eye, reminiscent of Joseph's interpretation of Pharaoh's dreams that safeguarded nations from famine. Or the operations manager who introduces AI-driven logistics to anticipate needs before they arise, embodying the virtue of foresight extolled throughout scripture.

With clarity and a commanding tone, I urge you, my dear reader, to envision your role in this brave new world. Integrate AI into your profession seamlessly, using discernment and wisdom to secure favor and influence. **AI is no longer the sole domain of engineers and developers; it is now a tool for the astute financier, the creative marketer, and the visionary manager to wield with purpose.**

Seek out these opportunities; the rewards for those ready to embrace AI are bountiful and promising.

AI, also known as artificial intelligence, is not confined to the realm of technology. Its applications extend far beyond the confines of coding and software development. Contrary to popular belief, AI is relevant across diverse fields, including content creation, marketing, sales, customer service, finance, and operations. Understanding the broad applications of AI beyond technology-related fields is crucial for professionals in various industries to recognize the potential it holds for unlocking new possibilities in their roles.

In content creation, AI can revolutionize the way information is generated, enabling automated content generation based on data analysis and user preferences. **The use of AI-powered content creation tools can significantly streamline the process of producing high-quality and relevant content, saving time and resources for businesses and creative professionals alike.**

Furthermore, the impact of AI in marketing cannot be overstated. AI empowers marketers to leverage predictive analytics and personalized targeting, resulting in more effective and impactful campaigns. **By analyzing vast amounts of data, AI can help marketers optimize their strategies, enhance customer engagement, and produce better advertising.**

In the realm of sales, AI can provide invaluable support by facilitating **lead scoring** and **predictive insights**. These capabilities enable sales professionals to prioritize prospects with the highest potential for conversion and tailor their approaches based on data-driven predictions, ultimately boosting their effectiveness in closing deals.

Customer service stands to benefit immensely from AI, particularly through the implementation of **chatbots** and **virtual assistants**. These AI-driven tools can deliver personalized assistance, resolve common queries, and provide real-time support, enhancing the overall customer experience while reducing the burden on human customer service representatives.

In finance, AI plays a pivotal role in **risk assessment**, **fraud detection**, and **algorithmic trading**. The ability of AI to process and analyze vast financial datasets with speed and precision makes it an invaluable asset for financial institutions and industry professionals seeking to make informed decisions and mitigate risks effectively.

Moreover, AI's relevance extends to operational functions, where it can optimize processes through **automation**, **predictive maintenance**, and **supply chain management**. By harnessing AI-powered solutions, operations professionals can streamline workflows, minimize downtime, and enhance the efficiency of their organizations.

Understanding the diverse applications of AI beyond technology-related fields is essential for professionals in various industries to harness its potential. By recognizing the breadth of opportunities AI presents, individuals can proactively explore ways to integrate AI into their roles and drive value creation in their respective fields.

Discover how AI can transform your professional landscape across diverse fields. **Read on to uncover the untapped potential of AI in content creation, marketing, sales, customer service, finance, and operations.**

AI is a transformative force that holds the potential to unlock new possibilities for value creation across various industries and professions. Its relevance extends far beyond the realms of technology, permeating fields such as content creation, marketing, sales, customer service, finance, and operations. By understanding how AI can be leveraged in diverse sectors, professionals can harness its power to elevate their roles and solidify their relevance in an increasingly digital world.

In **content creation**, AI tools can automate tasks such as content curation, writing, and editing. This can save time and resources for professionals while maintaining high standards of output. Similarly, in **marketing**, AI-powered data analytics can provide valuable insights into consumer behavior, helping professionals tailor their direct response advertising and brand-building strategies for maximum impact. By incorporating AI in **sales**, businesses can automate lead generation, optimize pricing strategies, and enhance customer relationship management.

Moreover, the integration of AI in **customer service** can elevate the quality of interactions, offering personalized support and streamlining processes. In the **finance** sector, AI algorithms can analyze large volumes of data at a speed and scale impossible for humans, enabling professionals to make more informed decisions and manage risks effectively. Additionally, AI can revolutionize **operations** by optimizing supply chain management, logistics, and resource allocation.

By recognizing the vast applications of AI beyond technology-related domains, professionals in diverse industries can unlock new realms of potential for value creation. This understanding enables them to navigate the increasingly digitized landscape with confidence, adapting their skills and practices to align with the evolving technological paradigm.

Understanding the pervasive impact of AI across various industries instills a sense of urgency in professionals, compelling them to stay at the forefront of technological advancements in their respective fields. The prospect of leveraging AI to transform practices, optimize processes, and redefine value creation presents an opportunity for professionals to not only adapt to change but also to lead it.

As we delve deeper into the potential of AI across diverse fields, it becomes evident that **the integration of AI is not merely an**

option but a necessity for professionals seeking to thrive in the digital age. Embracing the power of AI, professionals can forge a path towards unparalleled efficiency, innovation, and value creation within their teams, companies, and industries.

It is imperative for professionals across various domains to explore the applications of AI within their roles, seeking opportunities to integrate it into their practices and operations. Through proactive learning and adaptation, individuals can position themselves as indispensable assets within their organizations, harnessing the potential of AI to spearhead impactful change and advancement.

The transformative power of AI is not confined to technical domains; its potential is disseminated across myriad spheres, just as the wisdom of time-honored spiritual texts finds relevance in myriad aspects of life. In content creation, AI is akin to the proverbial lamp unto our feet, illuminating the path for writers and designers by proffering novel ways to craft narratives and visuals that engage and resonate. Content strategists now harness AI's capabilities to predict trends, personalize content to individual reader preferences, and even assist in creating written content with tools that suggest grammar improvements or generate topic ideas. Such tools are not replacements for the human touch in creativity but are valuable assistants that enhance productivity and innovation.

In the realm of marketing, AI stands as a beacon, guiding campaigns to higher levels of efficiency and personalization. To obtain a great return on investment, marketers must place their messages before the most receptive audiences for maximum impact. AI-driven analytics empower marketers to identify and target ideal customer segments, optimizing both ad placement and messaging based on data-driven insights. Thus, not only does this augment the potential for customer engagement, but it also ensures a prudent stewardship of resources, a principle esteemed in both business and spiritual teachings.

Sales professionals also find a powerful ally in AI. Like the astute servant entrusted with talents who invests them wisely, salespeople can leverage AI to glean insights from customer data and refine their sales strategies, ensuring they meet patrons at their points of need with precision, confidence, and empathy. The adoption of AI tools that automate routine tasks, analyze customer sentiment, and forecast purchasing trends allows sales teams to focus their efforts on fostering deeper human connections and closing deals with such balance that no machine can replicate on its own.

Customer service is another arena where AI redefines engagement. In this field, AI acts similar to the good shepherd, ensuring that every individual customer feels heard and cared for. Chatbots and virtual assistants, guided by AI, handle routine inquiries and transactions, freeing up human agents to tackle more complex customer needs with compassion and understanding. The ability to rapidly sift through vast datasets equips customer service representatives to provide personalized solutions swiftly and respectfully.

In finance, AI is the discerning advisor, able to analyze market trends, review quarterly earnings reports, manage risks, and provide recommendations with unprecedented speed and accuracy. Financial analysts can now utilize AI for tasks such as fraud detection and credit scoring, employing these intelligent systems as sentinels that watch over the integrity of transactions. Moreover, AI can assist in personalizing investment advice, drawing from vast pools of data to cater to individual investor goals and investment risk profiles. This accord of technology and personalization echoes the value of prudence and individual consideration championed in wisdom literature.

Operations management is yet another domain subject to AI's transformative influence. AI-driven supply chain management systems exemplify the principle of stewardship by ensuring resources are utilized judiciously, minimizing waste, and enhancing efficiency. Process automation optimizes workflow, predictive maintenance anticipates equipment issues before they occur, and demand forecasting ensures stock levels are kept in harmony with consumer needs. Thus, operations are not only streamlined but also infused with a foresight from both a productivity and profitability basis.

To identify opportunities to leverage AI is to assert one's agency in a rapidly evolving landscape, seizing the tools at our disposal to build careers and businesses that are not just sustainable, but also reflective of the profound respect for intellect and growth found in our shared human tradition. **Within each industry, each professional can find in AI a means to refine their craft, to serve their clientele better, and to manifest an enduring legacy of excellence and progress.** As we continue to navigate this bold new world of artificial intelligence, let it be with the resolve to harmonize innovation with the timeless principles that underpin our ethics and our professions.

AI's impact on diverse fields is undeniably profound, with the potential to revolutionize the way we work, create, and serve. As we delve into the myriad applications of AI, we must not lose sight of the spiritual principle that underpins our journey. **For in embracing new technologies and leveraging AI in our respective fields, we should honor the ethical call to be prudent stewards of the knowledge bestowed upon us.**

As we reflect on the possibilities unlocked by AI, we are reminded of a teaching that urges us to use our talents wisely and multiply them for the benefit of all. This is what AI offers – an opportunity to amplify our skills and expand our capacity to generate value for our organizations and communities.

The road ahead may seem daunting, but it is paved with promise and potential. As we traverse this landscape, let's keep in mind the words of ancient wisdom: *"As iron sharpens iron, so one person sharpens another." (Proverbs 27:17)* Let us sharpen one another's abilities and insights, learning from AI to refine our approaches, innovate our processes, and enhance our every endeavor.

In the realm of business and entrepreneurship, the unfolding AI frontier calls for courage, resolve and action. It beckons us to harness its capabilities and mold them in service of our ambitions. As we navigate the realms of content creation, marketing, sales, customer service, finance, and operations, let us seize the tools at hand and craft a future-worthy skill set.

In every field, AI presents an invitation to adapt, innovate, and excel. Its relevance is not confined to a select few; it permeates through diverse domains, offering invaluable prospects. By recognizing - and taking advantage of - these opportunities, we position ourselves not just to survive but to thrive in an AI-driven world.

Chapter Three

Navigating AI Ethics and Societal Impact

In the heart of Silicon Valley, as the sun painted the sky in hues of gold and orange, there was a gentle stir in Carter's home office. Amid shelves laden with literature ranging from Kierkegaard's existential musings to the latest on quantum computing, he sat before his terminal, contemplative. The warmth of the sunset was lost on him, for his thoughts sailed turbulent seas. The ethics of artificial intelligence —it was a problem as intricate and pervasive as the code he weaved.

Carter's foray into artificial intelligence was not solely a chase after technological prowess; it was a pilgrimage for a higher cause. He had long admired the prophets and philosophers who sought righteousness, and now facing a modern conundrum, he leaned into their wisdom for guidance. Theoretical frameworks were his Moses, leading him out of the wilderness of moral ambiguity.

Yet, the gravity of the situation weighed heavily. Each line of code was a step towards potential benevolence or unintended malevolence. Carter pondered on trust, that fragile bridge between innovation and acceptance. He knew that to cross it, he would need to balance the entrepreneurial spirit with the humility of a servant. As he gazed at the monitor, reflections of code scrolled like scripture, demanding not just his intellect but his conscience.

A chime of the doorbell snapped him from his reverie, a reminder of the mundane world beckoning for attention. It was a colleague, Ava, delivering a book on ethical dilemmas she promised would be enlightening. They exchanged brief pleasantries, yet Carter's mind was elsewhere—the intersection of AI and society, where the rubber met the road. Ava, sensing his preoccupation, offered a graceful smile before departing. There was work to be done, and he was the craftsman.

Returning to his desk, Carter rested his hands on the keyboard, still as if in prayer. He whispered a verse from Proverbs, seeking divine fortitude and guidance for the task ahead. His enterprise was not mere profit, but a testament to human ingenuity and integrity.

He strode forward, one block of code at a time, aspiring to engrave principles into each algorithm: transparency, accountability, fairness. This was no easy feat, yet the endeavor itself was fulfilling—a testament to his expertise and the legacy he yearned to leave. As the night drew its curtain, Carter embedded a promise within the digital framework, a covenant between creation and creator.

And so the question arose, whispered through the circuits and into the winds of change: In a world increasingly shaped by bytes and beliefs, how do we ensure the soul of humanity remains the compass?

The Moral Fabric of AI: Crafting a Conscience in Code

In our increasingly digital existence, professionals who ply their craft within the realms of artificial intelligence must consider their roles as stewards of a technology that inevitably molds society.. A vision for such new technology holds paramount importance, for without it, the very essence of our collective future risks being undermined by unintended consequences. To wield AI with wisdom and foresight, one must embrace a reverence for the ethical dimensions that govern its deployment, ensuring that progress aligns with principle.

Professionals navigating this landscape should strive not merely for compliance, but for the cultivation of trust and legitimacy in their endeavors. Just as a craftsman takes pride in their work, leaving no detail to chance, so must we meticulously **craft each interaction with AI to respect the dignity of all individuals it may affect**. Adopting a responsible and ethical approach is not only about mitigating risks but also about elevating the value we deliver through our professional contributions. In doing so, we build a foundation of integrity that transcends the immediate benefits of AI, reaching for a legacy of harmony and human-centric innovation.

Societal implications of AI span far and wide, echoing through the corridors of power, economics, and culture. It is an invitation to dialogue, drawing insights from various fields such as theology, politics, and economics. A well-rounded understanding in these areas enhances our capacity to navigate AI's complexities with acumen and empathy. **One must grapple with the profound questions that AI poses: How will it reshape labor markets? In what ways might it alter our conception of creativity and productivity?** Professionals poised at the helm of AI-enabled ventures have the responsibility to seek answers, ensuring that their contributions serve the common good.

The connection between ethical AI and professional success cannot be overstated. It creates a bond of trust with clients, users, and the broader community. When we predicate our careers upon the pillars of integrity and ethical consciousness, we establish a durable platform from which to launch into future opportunities. Ethical excellence becomes the distinguishing mark of our professional identity, attracting like-minded partnerships and fostering a network grounded on shared values. We must remember that in the eyes of faith, our work is a reflection of our inner selves, and as such, it should emanate the virtues we hold dear.

To engage with AI responsibly, professionals must cultivate keen insight and discernment. It is crucial to not only ask what AI can do but also to ponder what it should do. Educational initiatives and ongoing conversations within the field are necessary to shift the focus from mere functionality to foundational integrity. Empowerment lies in the ability to harmonize the sophistications of technology with the simplicity of moral clarity, ensuring our technologically driven actions never outpace our ethical compass.

In navigating AI ethics, we heed the proverbial wisdom that echoes across time, reminding us that true understanding stems from a blend of knowledge and reverence. **The professionals who emerge as leaders in the AI space will be those who integrate conscientious thoughtfulness into their practice.** Engaging with AI technologies requires a balanced blend of vigilance and vision, anchoring us in the present while steering us towards a future that upholds human dignity alongside technological marvel.

AI Ethics as a Professional Beacon

The pathway through the ethical dilemmas posed by AI is not one to tread lightly. It requires a heart aligned with virtue and a mind sharpened by discerning wisdom. As we immerse ourselves in the intricate world of AI, let us hold onto the values that have guided great leaders before us, allowing them to illuminate our path and inspire our actions. The societal trust we yearn to foster hinges on the choices we make today, and it is only through informed and ethical decision-making that we will forge a future where technology amplifies the best in humanity. Let us, therefore, pursue our goals with a guided conscience, anchoring our professional lives in the rich soil of ethical integrity.

Ethical considerations surrounding AI deployment are paramount, and professionals must engage with these technologies responsibly and ethically to align their actions with broader societal implications. **The impact of AI on society can be profound, touching on aspects such as privacy, bias, employment, and decision-making.** As professionals working with AI, it is crucial to understand the importance of ethical considerations in AI deployment.

AI technologies have the potential to transform the way businesses operate, solve complex problems, and enhance decision-making processes. However, the deployment of AI also brings forth ethical dilemmas that require careful consideration. Professionals must actively pursue a deep understanding of these ethical implications to ensure that their AI initiatives have a positive impact on society while upholding ethical standards.

Incorporating ethical considerations in AI deployment is not only a moral imperative but also a strategic necessity for professionals. It ensures that AI technologies are developed and utilized in a responsible manner, mitigating potential harm to individuals and communities. **By prioritizing ethical considerations, professionals can build trust with all stakeholders, foster a positive reputation, and contribute to a sustainable and ethical AI ecosystem.**

Professionals working with AI are in a unique position to influence the ethical development and deployment of AI technologies. The decisions they make regarding AI design, implementation, and governance can have far-reaching implications for individuals, organizations, and society at large. Therefore, a deep understanding of ethical considerations is crucial for navigating the complex landscape of AI deployment.

By embracing ethical considerations in AI deployment, professionals can help shape a future where AI technologies are used to enhance human well-being and promote fairness and equity. This involves actively engaging with ethical frameworks, codes of conduct, and industry best practices to ensure that AI is developed and deployed in a transparent, accountable, and inclusive manner.

As professionals, it is imperative to recognize that ethical considerations in AI deployment are not solely a matter of compliance but an opportunity to drive positive change. By actively addressing ethical concerns, professionals can contribute to an environment where AI technologies are harnessed for the collective good, empowering individuals, businesses, and communities.

It's time to delve deeper into how professionals can engage with AI responsibly and ethically.

As professionals engage with artificial intelligence (AI), it is crucial for them to do so responsibly and ethically. By aligning their actions with ethical principles, professionals can ensure that their use of AI contributes positively to the broader societal impact. This responsibility extends beyond personal gain and encompasses a commitment to upholding ethical standards that benefit the greater community. **By fostering an environment of trust and legitimacy through ethical engagement with AI, professionals can position themselves as leaders in their respective fields.**

One way for professionals to engage with AI responsibly is by prioritizing transparency. This involves being transparent about the capabilities and limitations of AI systems, especially when it comes to decision-making processes. Transparency builds trust and helps stakeholders understand the ethical considerations behind the use of AI. Additionally, it allows for informed consent and participation, which is essential for maintaining ethical standards.

Furthermore, professionals can engage with AI ethically by ensuring that the data used to train AI models is representative and free from biases. Biases in data can lead to discriminatory outcomes, impacting individuals and communities in negative ways. By addressing biases in AI systems, professionals can contribute to a more equitable and just society. Implementing measures to mitigate biases can include using diverse and inclusive datasets, as well as regularly evaluating and adjusting AI models to minimize potential bias.

Another key aspect of ethical engagement with AI is prioritizing the privacy and security of individuals' data. Professionals must uphold the privacy rights of individuals and protect their data from misuse or unauthorized access. This can be achieved through implementing robust data protection measures and adhering to established privacy regulations and standards. By prioritizing data privacy and security, professionals can reinforce trust and maintain respect for individuals' rights in the digital landscape.

Incorporating ethical considerations into the design and development of AI applications is also essential. Professionals should adopt an ethical-by-design approach, where ethical considerations are integrated into every stage of the AI development process. This involves evaluating the potential impact of AI systems on individuals and society, while also considering the long-term implications of their deployment. **By proactively addressing ethical concerns during the design phase, professionals can minimize the risk of unintended negative consequences**.

Moreover, **engaging with AI ethically involves actively seeking feedback and input from diverse stakeholders**. This includes individuals impacted by AI technologies, as well as experts in ethics, law, and social sciences. **By fostering open dialogue and collaboration, professionals can gain valuable insights into the ethical implications of AI and make informed decisions that consider a wide range of perspectives**.

Ultimately, engaging with AI responsibly and ethically requires a commitment to upholding moral and ethical principles in the application and deployment of AI technologies. **By integrating transparency, data fairness, privacy protection, ethical design, and inclusive collaboration into their AI endeavors, professionals can make a positive and lasting impact on the ethical landscape of AI deployment**.

The AI Integration Process Model

Setting Clear AI Objectives

Every impactful journey into the realm of AI begins with knowing the destination. Defining objectives is akin to setting the compass in the worldly pursuit of success with artificial intelligence. This step entails a profound contemplation on what organizations aspire to achieve through AI, grounded in both the tenacity of enterprise ambition and the prudence of ethical foresight. It is during this contemplation that the wisdom of spiritual texts can guide decision-makers to seek not only profit but also the greater good. By clearly articulating these goals, professionals take the first step not only toward advancing their business interests but also toward aligning their endeavors with the best interests of the communities and stakeholders they serve..

Assessing AI Readiness

Preparing the soil before sowing the seed is an age-old wisdom applicable to the deployment of AI. Assessing readiness involves introspection into the current state of technological infrastructure, data landscapes, and the prevailing organizational culture. Like the builders of old scrutinized blueprints before laying foundations, this assessment lays bare any potential stumbling blocks, cultivating the grounds for an AI integration that is both fruitful and well-conceived. When readiness coalesces with capability and intent, the result is a fertile ground for responsible and ethical AI deployment.

Crafting a Strategic AI Roadmap

Developing a strategy becomes the bridge between the aspiration of AI objectives and the reality of its enactment. It translates visionary goals into actionable plans, weaving the intrinsic value of technology with the fabric of the organization's mission. It involves a confluence of technological acuity, strategic vision, and a conscientious approach that reveres the wider impact of AI on society. As professionals chart their course, they must carry the lantern of responsibility alongside the torch of innovation, ensuring that every strategic move is illuminated by both.

The Rigor of Data Collection and Preparation

The collection and preparation of data is a pivotal act—scrutinizing the data's provenance, ensuring its integrity, and refining its quality are essential precursors to AI model development. Not unlike the meticulous craftsmanship of a potter shaping clay, this process molds the raw material of data into a form that's ready to breathe life into AI solutions. This thorough preparation of data ensures that it stands as a pillar of truth in AI's potential edifice.

Model Development: The Heart of AI

The core of any AI system is its model, developed through the selection of appropriate algorithms and meticulous training. This stage is where raw data metamorphoses into profound insights, driving intelligent decision-making. As the AI model is developed, professionals act as both artisans, finely tuning their creation, and as shepherds, guiding it to ensure that it is nurtured with the values of impartiality and fairness. The commitment to fostering models that uplift and empower rather than displace and discriminate becomes the mark of a professional who honors the societal contract.

Implementation and Integration

Following the creation of the AI models, their deployment is the act of bringing theory to practice. Like a symphony conductor ensures that each instrument plays in harmony, professionals must integrate these

AI models into the current systems seamlessly. This melding of art and science sees algorithms becoming part of the operational fabric, where their performance touches upon the everyday pulse of the business. It is in this practical symphony that the value of AI is made manifest, harmonizing efficiency with ethical use.

Monitoring, Evaluation, and the Cycle of Improvement

Perseverance in vigilance defines this stage. Once AI systems operate within the business landscape, monitoring and continuous evaluation ensure that AI's melody plays on key. Business outcomes become the measure of AI's true symphony, and through a disciplined process of setting performance indicators and adjustment, AI systems are fine-tuned to the dynamic rhythm of enterprise demands. Just as a gardener tends to a growing plant, monitoring and evaluation provide the feedback necessary to nurture and adapt AI systems to the ever-evolving needs of the business and society.

Refinement For Ethical Harmony

Evolving an AI system through refinement and improvement embodies the essence of stewardship in a technological world. It celebrates an organization's commitment to betterment, acknowledging that the first iteration is but a step on the path to excellence. Learning from the system's performance and the societal impact it wields, we engage in an iterative process—cultivating AI that serves the dual horizon of business growth and ethical integrity. This continuous cycle of improvement is not simply an act of refinement, but an expression of an enduring covenant with both our values and our vocation.

In encapsulating this framework, the primary objective remains steadfast: to navigate the implementation of AI with a scrupulous compass, aligning professional conduct with societal betterment. It is through this model that organizations can harness AI's transformative power while upholding a covenant of trust and ethical responsibility, propelling them toward a future where both business and society flourish under the wings of innovation.

In navigating the complex landscape of AI ethics and societal impact, it is essential to remember the timeless wisdom that reminds us to treat others as we would like to be treated. Whether considering the ethical implications of AI deployment or pondering the wider societal effects, these principles remain steadfast guides in the pursuit of ethical and responsible AI engagement.

As professionals engaged with AI technologies, **treating ethical considerations as a sacred duty rather than a burden can lead to more responsible and impactful decisions**. Embracing this perspective allows us to act from a place of integrity and reverence for the potential consequences of our actions.

When wrestling with the societal implications of AI and the role of professionals in fostering trust and legitimacy, the timeless teachings of social responsibility and justice from various religious and spiritual doctrines remind us to act in the service of the greater good. Embracing this call to action, working professionals can endeavor to uphold societal values and aspirations, and work towards contributing to the betterment of society through the ethical and responsible deployment of AI.

As we traverse the realm of AI ethics and societal impact, it is imperative not to lose sight of the interconnectedness of our actions with the greater fabric of society. Drawing upon ethical teachings and spiritual principles, professionals engaged with AI can harness the power of wisdom and compassion to navigate this terrain with reverence and purpose. By doing so, they can **serve as custodians of ethical AI deployment, paving the way for a future built upon trust, legitimacy, and societal well-being.**

Chapter Four

Mastering Data Literacy for AI Collaboration

B eneath the soft hum of the computer servers, Jenna leaned over her sleek desk, intently studying the charts and graphs illuminating her screen. The room was aglow with the glow of late afternoon—sunbeams sneaking past the blinds, casting strips of light that danced across her focused face.

She recounted advice from Proverbs 20:18, "Plans are established by seeking advice; so if you wage war, obtain guidance." Jenna was waging a war of a different kind—a war against the chaos of ungoverned data. Data had become more than mere figures; it was the beacon guiding her company through the treacherous and ever-changing seas of the market. In moments of solitude, Jenna often found her thoughts drifting to the countless businesses that failed to recognize the power held in their own numbers.

This particular problem struck her not only in its complexity but in its significance. She had to forecast sales for the coming quarter and the pressure was a significant force against her. The organization depended on her analysis to drive their strategies, to innovate, and to flourish. As she pondered on the scriptures, she found herself seeing data as David's stones against Goliath—small in appearance, but mighty in purpose.

From her years of tireless apprenticeship, she learned much. It was not just in her analytical skills, but in her faith and perseverance, echoing a sentiment found in Galatians 6:9, "Let us not become weary in doing good, for at the proper time we will reap a harvest if we do not give up." Jenna was a harvester of insights, cultivating strategies from the seeds of raw numbers. She felt the thrilling rush one gets when standing on the threshold of discovery.

Just then, the wind chimes sung a melodious tune as a breeze swept through her open window, ruffling notes and papers like a gentle reminder of the world outside these numbers. It whispered to her, a soft nudge not to be consumed by the quest for precision but to remember the why of her endeavor. Jenna was not merely dissecting figures. She was a steward of wisdom, shaping the future with careful, calculated strokes. Her entrepreneurial spirit did not just crave success; it was married to purpose, aligned with a faith that urged her to build not merely a company but a legacy.

As the day concluded, Jenna sat back in her chair, the room now bathed in the azure tinge of twilight. She exhaled deeply, the weight of data and decisions momentarily lifted. Tomorrow, she would wield this sword of information with the finesse and confidence of a master. The analytics were her canvas to paint — rich with patterns, telling a story only she could interpret.

At that quiet moment of reflection, as Jenna looked toward the night sky swathed in stars, a question lingered in her mind: How can the unwavering truths found in ancient wisdom illuminate the path through modern complexities, guiding us to decipher the stories held within the data?

The New Literacy Imperative

In an era where artificial intelligence (AI) remakes industry landscapes and reshapes how we approach work, **a new literacy is ascending as the cornerstone of professional competency**. It's not the ability to read texts or interpret numbers in isolation—**it's the mastery of data literacy intertwined with AI that has emerged as an indispensable skill. Those who can analyze, interpret, and visualize data, transform these analytical insights into strategic actions, and form dynamic alliances with AI are poised to thrive amidst the torrents of digital transformation.**

With data as the new currency and AI as its industrious custodian, a professional's prowess in navigating this relationship can define career trajectories. **It's no longer sufficient to delegate the realm of data to analysts or AI specialists, for the most impactful decisions now arise from a synergy between human intuition and machine intelligence. Gone are the days of data being an exclusive domain of technical experts. To add real value in any role or industry, understanding data is a shared prerogative, an essential thread in the fabric of AI collaboration.**

Empowering Growth Through Data Collaboration

To flourish, one must engage with data not as a passive observer but as an active participant. It's about developing a data mindset—one that sees beyond mere figures and charts to unearth the narrative within numbers. **It's about fostering an analytical acuity to discern patterns and make predictions, a critical eye that questions the source and relevance of data, and a creative vision that crafts compelling stories through data visualization.**

The mark of true data literacy extends beyond personal mastery; it signifies one's capacity to be an AI collaborator—where one's insights guide AI towards more nuanced analyses and where AI becomes an extension of one's analytical powers. **This symbiosis is the bedrock of informed decisions, effective problem-solving, and the evolution of strategic initiatives that separate the proficient from the novices.**

Translating Complexity into Clarity

Imagine facing a labyrinthine array of data points and without apprehension, steadily extracting clarity and meaning. Imagine harnessing predictive models that afford a glimpse into future trends or averted crises. This is the transformative power of data literacy—**transforming complex datasets into actionable insights**, demystifying the abstract to arrive at concrete strategies that bolster business objectives and forge innovative solutions. **Data literacy is no longer an elective; it's imperative**. It's the tool that dismantles complexity, enabling businesses and professionals to tread confidently into data-rich environments, hand in hand with AI.

The AI Synergy Blueprint

Navigating the AI landscape can seem daunting. Yet, each professional's journey to AI synergy is navigable with a clear blueprint—a step-by-step process that guides you from awareness to mastery, from potential to actualization. Embarking on this journey requires methodical evaluation and tailored strategies. It starts with recognizing the role of AI in your specific field and extends to enhancing your skillset to meaningfully engage with AI-powered tools and insights.

Step 1: Identifying the Role of Data and AI in Your Sphere

Every profession has unique datelines—an intersecting point where data and AI meet the practical demands of your job. In this first step, you'll learn to pinpoint where these lines cross in your field and the implications they hold for your role. This foundational understanding is paramount for building upon the subsequent steps.

Step 2: The Research Reconnaissance

Immerse yourself in the realm of AI applications relevant to your niche. Information is power, and power lies in the knowledge of AI's current and potential impact. Let case studies become your reference, articles your roadmap, and industry reports your compass as you navigate through the sea of AI possibilities for your role and industry.

Step 3: Self-Audit for AI Readiness

Turn the lens within—evaluate your current capabilities regarding AI. Determine which skills are robust, which require honing, and which remain to be cultivated. This **self-reflection** sets a baseline for growth as you align your professional repertoire with AI's possibilities.

Step 4: The Quest for Knowledge Enhancement

Armed with insight from your self-audit, pursue avenues for continuous learning and improvement. Identify courses, books, videos, workshops, or certifications pivotal in bridging the gap between your current skill set and the AI proficiency required. This isn't mere education—it's a strategic investment in your professional future.

Step 5: Strategy Tailored to Your Trajectory

Drawing from the intelligence gathered, concoct a personalized strategy. Set goals, choose technologies of focus, and outline a sequence of actions leading to your upskilling in AI—the precision of this strategy will reflect in the sharpness of your professional edge, especially among your colleagues.

Step 6: The Vigilant Execution and Evolution

Implement your strategy diligently. As you do, retain agility to calibrate your approach in response to the evolving AI landscape. Your adaptability unveils new vistas of opportunity and emboldens your professional resilience.

By ardently pursuing data literacy and molding yourself into an AI collaborator, your career becomes not just future-proof, but dynamically poised to lead and triumph amidst the ceaseless waves of technological innovation. The harmony between data insight and AI's prowess is a duet that, once mastered, plays the anthem of modern professional success. **Engage with this chapter as if your career depends on it—because, in many ways, it does.**

In the age of artificial intelligence (AI), developing proficiency in data analysis, interpretation, and visualization is essential for professionals seeking to collaborate effectively with AI technology. Mastering data literacy not only enables individuals to understand complex datasets but also empowers them to harness the power of data-driven insights for informed decision-making and problem-solving. This proficiency positions professionals as adept collaborators with AI and equips them to translate intricate datasets into actionable strategies and solutions.

Data Analysis: Developing proficiency in data analysis involves the ability to examine large volumes of data to uncover hidden patterns, correlations, and trends. This skill enables professionals to derive meaningful insights from datasets and make informed decisions based on evidence rather than intuition.

Data Interpretation: Interpreting data involves understanding the significance of the patterns and trends identified through analysis. It requires professionals to draw conclusions from the data and use these insights to inform decision-making processes in various domains, from business strategy to marketing campaigns.

Data Visualization: The ability to communicate complex data in a visually compelling manner is crucial for professionals working with AI. Data visualization involves presenting data in charts, graphs, and interactive dashboards, making it easier for stakeholders to understand and act upon the insights derived from the data analysis.

Proficiency in data analysis, interpretation, and visualization isn't just about understanding numbers; it's about translating raw data into meaningful, actionable insights. By honing these skills, professionals position themselves as indispensable collaborators with AI, able to leverage the power of data to drive informed decision-making and strategic planning.

Given the accelerating pace of technological advancement, mastering data literacy is foundational for professionals seeking to future-proof their careers. The ability to analyze, interpret, and visualize data not only prepares individuals to collaborate effectively with AI but also fortifies their expertise as valuable contributors to any organization.

In the next section, we'll explore how to harness the power of data-driven insights for informed decision-making and problem-solving in partnership with AI. Keep reading to discover how to leverage the full potential of data literacy in our AI-driven world.

Data-driven insights are at the heart of informed decision-making and problem-solving, serving as valuable guides in navigating complexities and uncertainties. Harnessing the power of data to glean actionable insights is a critical skill in today's data-driven world. Professionals who can leverage data to inform their decisions possess a unique advantage, enabling them to make strategic choices with confidence. By mastering data literacy, individuals can unlock the potential of data-driven insights to drive informed decision-making and problem-solving, positioning themselves as invaluable collaborators with AI.

Empowering Insights Through Data: Developing proficiency in data analysis and interpretation empowers professionals to uncover meaningful insights from complex datasets. Drawing upon the principles of data literacy and analysis, individuals can extract valuable information that serves as the foundation for informed decision-making. In a business context, these insights can shape strategies, optimize processes, and drive innovation, ultimately fostering growth and success.

Leveraging Insights for Informed Decision-Making: The ability to harness data-driven insights equips professionals with the tools to make informed decisions. Leveraging these insights allows individuals to assess risks, identify opportunities, and make strategic choices that are rooted in evidence and analysis. It provides a framework for evaluating scenarios, mitigating uncertainties, and charting an informed path forward.

Navigating Complexities through Data: When confronted with complex challenges, data-driven insights serve as a compass, guiding professionals through intricacies and uncertainties. Whether addressing market dynamics, customer behavior, or operational efficiencies, the ability to lean on data-driven insights fosters clarity and precision in decision-making, enhancing the likelihood of successful outcomes.

Empowering Problem-Solving with Data: In the realm of problem-solving, data-driven insights offer a wealth of possibilities. Understanding and utilizing valuable data insights enables individuals to identify the root causes of issues, design targeted solutions, and optimize outcomes. It serves as a cornerstone for strategic problem-solving, allowing professionals to address challenges with confidence and precision.

The Collaborative Power of Data-Driven Insights: Proficiency in harnessing data-driven insights not only empowers individuals but also paves the way for fruitful collaboration with AI. By mastering data literacy, professionals position themselves as adept collaborators with AI, ensuring that data-driven insights are seamlessly integrated into the collaborative process, thereby enhancing AI's ability to augment and optimize decision-making and problem-solving.

A Call to Action: Embracing the power of data-driven insights is not merely a professional pursuit, but a transformative journey that empowers individuals to make confident, informed decisions. By investing in data literacy and analysis, professionals can unlock the potential of data-driven insights, positioning themselves as astute collaborators with AI and catalysts for progress and innovation.

In the intersection where human discernment meets artificial intelligence, one finds the potential to extract profound value from data. It takes a nuanced approach to transform this data into strategies that can energize your enterprise. This is not merely about reading numbers and charts; it's about storytelling supported with data, where the narrative is driven by insight and propelled by AI's capabilities. Consider the parable of the talents, for instance. The wise are those who invest their resources wisely, yielding an increase. Similarly, utilizing AI for data analysis is an investment in knowledge, with the dividends paid in strategic opportunities and smarter business moves.

Take, for instance, a retail business analyzing customer patterns. Human analysts might easily become overwhelmed by the sheer volume of transactions. However, when in partnership with AI, these patterns emerge with clarity. AI can crunch numbers at an astonishing rate, identifying trends and anomalies alike. But it's the professional, equipped with both data literacy and a creative mind, who will interpret these patterns. They are the ones who will discern whether a spike in sales was due to a one-off event or indicative of a longer-term trend that should inform inventory decisions.

Professionals in this new era will benefit from adopting a mindset reminiscent of Biblical Joseph in Egypt, who interpreted dreams to forecast and prepare for years of both plenty and famine. With technological tools at your disposal, garner the foresight to anticipate market shifts. It's less about predicting the future and more about creating a resilient strategy that can weather various scenarios. AI facilitates scenario planning, crunching vast datasets to simulate outcomes, and equipping you with the wisdom to select the best course of action.

The sophistication of AI also means that it can model complex human behaviors. In marketing, this capability allows you to segment audiences with precision, crafting messages that resonate on a near-personal level. Yet, without understanding the cultural, emotional, and psychological drivers behind the data, messages may fall flat. It's the human touch, informed by data literacy, that refines these messages, ensuring that your outreach initiatives strike a chord with empathy towards the intended audience.

Harnessing the power of AI also means embracing continuous learning. Just as economies and industries evolve, so too must our strategies and solutions. AI offers not only a snapshot of the present but also a learning curve that iterates and improves over time. This concept echoes the timeless wisdom: "Give a man a fish, and you feed him for a day; teach a man to fish, and you feed him for a lifetime." By learning from AI, you are not just benefiting from a single insight but developing a skillset that can generate insights perpetually for the rest of your career.

Empowering your strategies with AI requires a leadership style that champions innovation and experimentation. In biblical terms, it's akin to David's boldness to face Goliath—not conforming to conventional armor but instead selecting five smooth stones from the brook. In your professional life, AI is that unconventional tool that, when wielded with expertise, can overcome the towering challenges in data complexity. Be bold and innovate; let AI assist you, but never forget that it's your unique human insight that guides its purpose and applications.

Ultimately, the cooperative endeavor between you and AI is about building a symbiotic relationship where each entity plays to its strengths. This collaboration isn't about one surpassing the other, but about complementing capabilities. As Proverbs 27:17 says, "As iron sharpens iron, so one person sharpens another." AI sharpens your ability to process and analyze data, while you sharpen AI's focus, directing its computational power to yield the most impactful insights.

In leveraging this relationship, remember that empathy and understanding are your guiding lights in the sometimes overwhelming world of data. AI can open the door to profound insights, but it is up to you to step through, interpret the information contextually, and act upon it judiciously. As a professional today, take on the mantle of interpreter and strategist, moving beyond basic data literacy to a place where, with the support of AI, you turn data into dialogue, numbers into narratives, and insights into strategies and tactics towards realizing a resilient vision for the future.

As we conclude this chapter, let's reflect on the wisdom of ancient texts that remind us of the value of mastering data literacy for our professional journey. In the words of Proverbs 24:3-4, "By wisdom a house is built, and through understanding it is established; through knowledge its rooms are filled with rare and beautiful treasures."

Proficiency in data analysis, interpretation, and visualization is the foundation upon which we build our professional "house." This proficiency equips us with the understanding to establish ourselves amidst the ever-evolving landscape of AI-driven collaboration.

Harnessing the power of data-driven insights empowers us to make informed decisions and solve complex problems. Such wisdom guides us toward decisions that not only benefit ourselves but also our teams, businesses, and the broader community.

Translating complex datasets into actionable strategies and solutions in collaboration with AI positions us as adept and influential collaborators. By translating data complexities into tangible strategies and solutions, we build our professional "house" on a solid foundation, ensuring its resilience and enduring value.

So, as we continue this journey together, let these teachings and timeless wisdom guide our pursuits, emboldening us to master data, embrace AI collaboration, and secure our professional futures. Embrace these insights as you move forward, and may they provide certainty and strength on your path to success. Let us press on in our pursuit to become masters of AI collaboration, leveraging data literacy with conviction and purpose.

Chapter Five

Cultivating a Growth Mindset in an AI-Driven World

I n the heart of the city, as the day yielded to dusk, Alex lingered in the warm glow of the office, pondering upon the grand tapestry of technology and its inevitable march forward. She could hear the placid hum of the air conditioning system, feel the rhythmic tap-tap of fingers on keyboards as her colleagues toiled alongside her. They were like monks in a modern monastery, devoted to the digital deities of code, prompts and algorithms.

Her desk, cluttered with papers strewn like fallen leaves in autumn, bore testament to the restless spirit of enterprise. On the screen before her, a window into another realm, artificial intelligence beckoned like a siren's call. Alex knew, deep within the chambers of her heart, that to stand still was to retreat. With each passing moment, she felt an urgency akin to the faithful heeding a divine summons, an urgency to evolve, to adapt, and to grow.

She had seen businesses born and later buried in the sands of complacency with their epitaphs written in the language of those who feared change. But not her; no, she was the daughter of a different creed. Her spirit, honed by challenges experienced through previous instances of creative destruction, coveted the sharpness acquired through the relentless pursuit of knowledge to stay ahead of the trend.

Alex's thoughts were interrupted by the gentle chime of her smartphone. She received a message from a mentor reminding her of the *parable of the talents* (Matthew 25:14-30). The message seemed to mirror her inner contemplation. "To those who have, more will be given," her Bible read, a present-day parable parallel to the rise of AI. It was a call to action, to invest the talents she had been blessed with, not to bury them in fear or in rebellion.

As the city lights began to twinkle like a celestial map charting unknown futures, Alex's resolve deepened. She reflected on a verse from Ecclesiastes 3:1-8 that spoke of the seasons of life, realizing this was her season to plant seeds to further the growth of her career. It was as though each line of code she understood, each algorithm she mastered, and each prompt she wrote was a step on the path to a promised land of innovation and increased job security.

Would she falter, she wondered, as the uncharted terrain of AI integration stretched before her? Or would she, with the vigor of David facing Goliath, slay the doubts that threatened to be her undoing? The power of the sling in her story lies in her perspective, her eagerness to learn, and her unwavering faith in the journey.

A soft sigh escaped her lips, a whisper of acknowledgment to the Creator who had crafted the human mind capable of fashioning such wonders as AI. As she prepared to leave the office, the ghost of a smile played upon her lips; for she knew that, though the road may be uncertain and arduous, it was also worthwhile. And isn't it in the sanctity of our toils that we find our true calling?

The Imperative of Adaptability

As we delve into the complexities of a rapidly evolving technological landscape, **it is abundantly clear that a growth mindset is not just a nice-to-have, but a vital necessity for professionals striving to maintain relevancy.** Proverbs 18:15 states, *"An intelligent heart acquires knowledge, and the ear of the wise seeks knowledge."* In an AI-driven world, these words ring truer than ever, underscoring the importance of seeking wisdom and knowledge continually. The burgeoning dominance of artificial intelligence in the workplace demands that we not only embrace change but also actively seek ways to integrate it into our personal and professional development.

The Curiosity That Fuels Progress

Adopting an attitude of curiosity in the face of AI is akin to cultivating a garden rich with diverse flora; it requires dedication, patience, and a genuine interest in the potential for growth. Proverbs 1:5 says, *"Let the wise hear and increase in learning, and the one who understands obtain guidance."* Applying this wisdom, we should approach AI not as a threat but as an invaluable source of skills and knowledge enhancement. By doing so, we can transform how we execute tasks and continue to add value to our roles, teams, and careers.

The technology landscape is shifting beneath our feet and right before our eyes, and **adaptability** is the solid ground we must seek.

In the business realm and beyond, it is imperative to understand the significance of being agile and resilient. By fostering a mindset that not only accepts but anticipates change, professionals of all industries can open doors to new opportunities that may have otherwise remained unexplored. This proactive approach to integrating AI into every facet of our professional lives illuminates pathways to innovation, efficiency, and, ultimately, career security.

The Proactive Spirit in the Age of AI

To remain proactive, one must be diligent in seeking knowledge. Similar to the *Parable of the Talents*, we are called to invest our abilities and reap the rewards of our labor and stewardship. AI is a tool—one that we must learn to wield expertly to achieve success in an age where stagnation equates to obsolescence. **Be like the servant who doubles their talents, not the one who buries them.** Aim to consistently hone your skills and broaden your knowledge base in the realm of AI, ensuring that you are an asset to your field and an architect of your own future.

In practice, a growth mindset demands both humility and confidence. It requires acknowledging the vastness of what we don't know while possessing the self-belief, courage, and persistence to master new concepts. In this chapter, we will explore not just why continuous learning and adaptability are vital, but how to embody these principles. We will address strategies to stoke your inherent curiosity and how to translate that into tangible, marketable proficiencies in an AI-augmented business environment.

Steering Through the Waves of Change

Indeed, the integration of AI in our professional lives can seem like navigating a ship through uncertain waters—with storms on the horizon. But just as a skilled sailor understands the winds and currents to use to their advantage, a savvy professional must learn to harness the capabilities of AI. This entails not only keeping abreast of technological breakthroughs but also actively seeking out training programs, networking opportunities and other resources to reinforce one's position at the helm.

It is a journey, a continuous process of self-improvement, and collective growth. As career professionals, we must transform anxiety in the face of the unknown into excitement for potential discoveries. This chapter is a call to action, an invitation to step boldly into the future and to craft a career that is not just sustainable but flourishes in the embrace of AI.

In today's rapidly evolving landscape, embracing a mindset of continuous learning and adaptability is crucial for professionals. Technological advancement, particularly the prevalence of AI, demands a proactive approach to personal and professional development. Instead of resisting change, professionals must be willing to learn, grow, and adapt. This growth mindset is not just a valuable asset; it is essential for thriving in an AI-driven world.

Embracing a growth mindset involves recognizing that one's abilities and intelligence can be developed, rather than being fixed traits. It requires a willingness to learn from mistakes, seek feedback, and consistently improve. In the context of AI, this mindset is particularly important, as it encourages professionals to approach new technologies with curiosity and enthusiasm, rather than apprehension.

The Value of Continuous Learning

Recognizing the value of continuous learning is the first step toward cultivating a growth mindset. It involves **acknowledging that the skills and knowledge relevant today may not be as valuable tomorrow**. The rapid advancement of AI and automation means that professionals who are stagnant in their knowledge and skills risk falling behind. By committing to ongoing learning and development, individuals position themselves to adapt to the changing demands of the digital era.

Mid-Career Learning

Especially seasoned professionals should be open to the idea of continually expanding their skill set and knowledge base. Mid-career learning is becoming increasingly important in the face of rapidly evolving technology. The ability to adapt and learn new skills will differentiate those who thrive from those who stagnate. That can mean the difference between receiving the promotion or not. By maintaining a mindset of continuous learning, professionals can ensure that they remain competitive and valuable in the job market.

Adaptability as a Competitive Advantage

In a world where the job market is continually shaped by technological advancement, adaptability is a competitive advantage. Professionals who are flexible, open-minded, and eager to learn new technologies are more likely to succeed in the face of automation. Embracing change and learning to work alongside AI rather than against it can lead to valuable insights, new opportunities and career growth.

The Ethical AI Integration Framework

In an age where artificial intelligence permeates nearly every aspect of professional life, fostering a proactive approach to personal and professional development is paramount. The Ethical AI Integration Framework (EAIF) offers a structured pathway for individuals and organizations to responsibly navigate the complexities of AI integration.

Identify Stakeholders

The identification of stakeholders forms the initial pillar of the EAIF. As moral agents, it's imperative to discern who will be impacted by the deployment of AI systems. Incorporating the perspectives of employees, customers, shareholders, and society ensures a holistic evaluation. **Key to this is recognizing that every decision has a ripple effect**, with potential to touch the lives of many, for good and for bad.

Define Ethical Principles

Defining ethical principles is the moral compass of AI integration. Such principles will anchor us in a sea of rapid innovation. Whether it focuses on privacy, fairness, transparency, and / or accountability, these agreed-upon tenets must mirror the ethical convictions we uphold in the broader spectrum of our lives and the organizations we represent. They serve as a beacon for businesses navigating the intricate moral landscape of AI, reinforcing the notion that technology should serve humanity in ways that align with our core values and mission.

Assess Potential Risks

Once stakeholders are identified and ethical principles established, assessing potential risks is a crucial step. This involves a vigilant analysis of both the immediate and the distant horizon of AI's impact. From potential data bias to job displacement, these risks are not merely hypothetical concerns but real-world issues that call for preemptive contemplation, strategic planning and resolution. **By assessing risks thoroughly, we prepare ourselves not just for probable outcomes, but also for the unforeseen**, much like the wisdom contained in the parable of the wise builder who lays a solid foundation (Luke 14:28).

Mitigation Strategies

Developing effective mitigation strategies is an exercise in stewardship and wisdom. It involves devising measures that counterbalance the identified risks, thereby nurturing an environment where ethical AI usage thrives. Here, we echo the principle of taking corrective action before missteps transform into stumbling blocks. This could manifest through rigorous algorithmic checks, employee education, and refining processes that uphold integrity and company ethics. Effectively, these strategies are akin to the ethical safeguards that keep operations in alignment with human dignity and respect.

Compliance and Governance

Compliance and governance represent the structural bedrock of ethical AI use. Establishing clear guidelines and policies ensures AI technology operates within a defined ethical and legal framework. As such, internal governance structures are formed, not as mere regulatory hoops but as embodiments of our commitment to moral righteousness and ethical responsibility.

Continuous Monitoring and Evaluation

Finally, the EAIF champions continuous monitoring and evaluation as an integral aspect of ethical AI integration. It is a dynamic, living process—reflective of the continual growth individuals are called to pursue. By regularly reviewing ethical guidelines and monitoring AI performance, organizations can remain responsive to feedback and make adjustments that resonate with stakeholder needs, especially as technology continues to evolve.

The EAIF model serves as a solid, multi-faceted guide for navigating the ethical implications of artificial intelligence, resonating with moral and pragmatic dimensions. It underscores the imperative of a growth mindset, encouraging individuals and organizations to approach AI with open minds and ethical vigilance.

As we implement these components, they harmonize together to form a symphony of responsible innovation, safeguarding the human spirit at the core of technological advancement. This model not only prepares for the present but lays the groundwork for a future where AI serves the greater good, anchored in ethical purpose and human-centric design.

As we further navigate the AI-driven world, let us remember that a growth mindset is not only a professional necessity but a personal asset. Continuously seek opportunities to learn, approach new challenges with excitement, and take proactive steps to enhance your skills. In doing so, you will undoubtedly thrive in the ever-evolving professional realm of AI integration.

Chapter Six

Building Resilience Amid AI's Impact on Careers

A midst the clamor of the marketplace, where the aroma of freshly ground coffee beans merged with the zest of haggling voices, Paul stood before the worn facade of his family's bookstore. It was an oasis of knowledge in a desert of ever-shifting digital sands. The city was a maelstrom of change, every day bringing news of another local store shuttering, succumbing to the relentless march of e-commerce giants and the onslaught of oncoming artificial intelligence-driven services. In this teeming bazaar of commerce and algorithms, Paul felt the tremor of impending change like a dissonant chord.

Inside the bookstore, the air was thick with the musty smell of paper and ink, a scent that was both comforting and, increasingly, an anachronism. Paul wandered through aisles of towering bookshelves, fingers tracing the spines of ancient tomes and modern bestsellers alike. This place was his legacy, a testament to generations of curators of thought. Yet, it was at the crossroads of transformation. The writing was on the virtual wall: adapt or become extinct.

He paused before a volume titled "The Resilient Spirit: Strategies for Thriving in Times of Change", its cover worn from frequent handling. His contemplation was deep, wrestling with the paradox of preserving tradition while staying afloat in a sea of innovation. He thought of his grandfather, who had established this haven with little more than a fervent belief, a thousand dollars, and a suitcase of books. His resilience had been his faith and his unshakeable conviction that knowledge was the divine light guiding humanity.

As he pondered, a customer approached, her request became a wake-up call to reality: she sought a digital version of a classic literary work, one not available in the hardcover clutched in her hand. It was a sign of the times, a clarion call that beckoned Paul to pivot and upskill, but how? Could his business embrace the digital era without losing its soul? Should he take a leap of faith, steering his ship into the unknown digital waters, trusting that the same resilience that anchored his grandfather would sustain him?

The sound of a child's laughter, echoing through the bookstore's quiet corners, reaffirmed Paul's conviction that his journey was not just about survival but also the enrichment of the human spirit. In the intersection of tradition and innovation, there was opportunity. The challenge before him was not an end but a beginning, if he so chose to participate in it, one that required the wisdom of Solomon and the adaptability of a chameleon.

And in the quiet sanctuary of his bookstore, Paul resolved to find the balance. He would learn, adapt, and integrate new technologies, but his mission would remain fixed—to kindle the flame of knowledge and imagination in the hearts of his patrons. It was a hopeful synthesis, a testament to the enduring power of faith and resolve.

Was his decision an omen of a renaissance for the small mom-and-pop entrepreneur in the digital age, or an idealistic delusion in the face of relentless progress? Only time, with its unerring and unyielding march, would tell.

The Bedrock of Adaptability: Building Your Resilience Fortress

In an age where artificial intelligence reshapes the contours of the workforce, the concept of career resilience emerges not as a luxury, but as a necessity. This profound shift beckons a timely response: the onus on professionals to adapt, upskill, and pivot in the face of rapid technological change. Such agency paves the way for career longevity and success in an era of flux. It's more than mere survival—it's about thriving in the AI-augmented landscape, **where *human ingenuity* meets *machine efficiency***.

The transformative power of AI beckons us to reimagine our career paths. As stewards of our professional destinies, we must embrace the challenge of continuously learning and relearning skills. This striving is both a safeguard against the tides of job insecurity and a solemn act of self-empowerment. For as we embrace the unknown with faith and foresight, we anchor ourselves in the wisdom that uncertainty can be navigated with grace and confidence.

In this pursuit, not only must we recognize the inevitability of change, but also view it through a lens of opportunity. The strategies outlined herein provide a roadmap for navigating this transforming ecosystem. By remaining vigilant and proactive, we can ensure not only our relevance but our ascent in an AI-driven economy. The entrepreneurial spirit flourishes most brightly when it is kindled by resilience and the unwavering resolve to rise above challenges.

The steps we can undertake to integrate AI into our work routines epitomize this ethos of resilience. They offer a concrete manifestation of our commitment to self-improvement and innovation. This pragmatic approach to AI adoption positions us at the vanguard of the new industry dynamics, allowing us not only to cope with the winds of change but to harness their energy for our career advancement.

Empowerment through Integration: Syncing Seamless with AI

Step 1: Identify Tasks or Processes for Automation

The first stride on our track to resilience involves discerning which work aspects can be elevated through AI. Pinpoint repetitive tasks or systematic processes that can be refined or expedited with the aid of AI technologies and tools. Cataloging these tasks provides a clear target for enhancement.

Step 2: Research AI Solutions

With your list of tasks and processes in hand, delve into the market to unearth AI tools tailored to these functions. Pore over reviews, analyze features, sign up for demos, and seek insights from peers or authorities in the field. This inquiry is vital in matching the right AI solution to your specific situation and challenges.

Step 3: Evaluate Potential Solutions

At this juncture, it's imperative to weigh these AI contenders against each other and your unique needs. Considerations should be broad, including user-friendliness, integration ease, cost implications, and the availability of ongoing support. Let these parameters steer you toward the most promising options.

Step 4: Test and Pilot

Prior to wholesale adoption by you and your company, a (free) trial run of your chosen AI tool is essential. Deploy it on a smaller scale or within a limited scope to gauge its effectiveness and gather real-world feedback—this pilot test is a critical step in ensuring the tool's alignment with your expectations.

Step 5: Implement and Train

Successful piloting gives way to full-scale implementation. Instruct and acclimate yourself and your team in the nuances of the chosen AI application, reinforcing its advantages and potential constraints. Mastery of the technology burgeons into organizational empowerment as you duplicate your know-how of the AI application to other team members to deepen its use, adoption and benefit for your organization.

Step 6: Monitor and Adjust

As the AI solution becomes woven into the fabric of your daily work, vigilantly oversee its operation. Monitor its efficacy, recognize and rectify shortcomings, and keep abreast of updates or advancements from the provider. This continuous attention solidifies the union between your work and AI. Rinse and repeat these steps with each AI tool.

Herein, we have sketched the contours of a modus operandi to ride the wave of AI—a wave that promises to surge through the vast ocean of career opportunities. Allow these strategies of adapting, upskilling and pivoting to be both your compass and your anchor, as you chart a path of professional resilience and lifelong learning to excel in today's fast-paced and ever-changing world.

Proactive Engagement with Technology

To effectively adapt, upskill, and pivot, professionals must proactively engage with technology, including AI and machine learning. This involves not only acquiring technical skills but also developing a deep understanding of how technology functions within specific industry contexts. By integrating technology into their skill set, professionals can position themselves as indispensable contributors within their respective fields.

Seeking Mentorship and Guidance

In the pursuit of adaptability, upskilling, and pivoting, seeking mentorship and guidance can be invaluable. Learning from those with expertise and experience in navigating the impact of newer technology on careers can provide critical insights and strategies for success. Mentorship offers the opportunity to gain wisdom from those who have traversed similar professional challenges, offering invaluable advice and support during uncertain times.

Building relationships with mentors, colleagues, and industry experts can provide the crucial support needed as careers transform. It is through these partnerships that we gain insight, encouragement, and opportunities for collaboration, which can open doors to new endeavors in an AI-driven economy.

At the same time, look for ways that you can help someone else in the process as a mentor to him or her. Offering actionable advice, grounded in firsthand experience, can equip individuals with practical tools to navigate the evolving landscape of their industries and guard against job insecurity. Pay it forward to those you know going through something similar in their careers.

Embracing the Unknown

The journey of adaptation, upskilling, and pivoting demands a willingness to embrace the unknown. It requires professionals to step into uncharted territory, confronting uncertainty with a sense of purpose and resolve. Embracing the unknown is a testament to one's resilience in the face of change, demonstrating a readiness to confront the challenges posed by AI's influence on careers.

Learn from Outside of Your Box (or Cubicle)

A broad range of insights from various fields can contribute to a well-rounded approach to cultivating resilience. By incorporating references and concepts from different disciplines, individuals can gain a rich, informative understanding of the challenges and opportunities presented by AI's impact on careers. Case studies, analogies, and diverse perspectives can provide valuable context and insight crucial for navigating the complexities of the professional landscape.

Today's professionals must proactively seek out the knowledge that will keep them competitive in a marketplace augmented by AI. This may involve regularly reviewing new industry trends, engaging in continuous professional development, acquiring certifications in emerging technologies, and / or learning to work symbiotically with artificial intelligence tools. This could take the form of online courses, workshops, or self-study, always with a focus on staying current with the latest advancements and understanding how they impact their roles, teams, companies, and industries.

Who knows? Just as Esther was placed "for such a time as this" (Esther 4:14), we might find that our adaptability and resilience amidst AI's disruption is positioning us for an influential role or once-in-a-lifetime opportunity we could not have previously imagined.

Chapter Seven

Leveraging Networking and Mentorship in the AI Age

Winter's chill had settled over the city, a biting reminder that even the steel and glass spires were not immune to nature's whims. Anthony moved through the downtown throngs, a man seemingly apart from the crowd, his breath a mist that mingled with the industrial exhalation of the metropolis. He walked with purpose, bound for a meeting that might well pivot the course of his burgeoning technology startup.

The café was bustling, a hub of conversation and commerce, a fertile ground for the ideas Anthony needed to cultivate. As he sat with his black coffee—no sugar, no cream—he thought of Proverbs 15:22: *"Plans fail for lack of counsel, but with many advisers they succeed."*

Anthony needed more than just a plan; he required the vision that only the experienced can enlighten.

He had arranged this meeting with Natalie, a guru in the AI space. As he awaited her arrival, he recalled his early fascination with artificial intelligence, the dreams that bounced wildly in his eager, youthful mind. He thought of the dreams that had become goals, and the goals that now verged on tangible realities. Yet, he was gripped by the weight of his aspirations, aware of the many who had faltered in the attempt to bring dreams into the light of day and the hegemony of the marketplace.

The door's chime announced Natalie's arrival; the woman's presence was a comfort, her bearing an assurance. Anthony greeted her with familiarity, yet with the deference due to a mentor. As they spoke, Anthony found solace in Natalie's words, proverbs in their own right, tailored to the challenges of AI integration. Natalie's anecdotes were pebbles of wisdom gleaned from a career spent weaving through complexities that lay like traps for the unwary.

"Remember, Anthony," Natalie intoned, "it's not simply about understanding the technology. It's about understanding people—how they think, work, and interact with the systems we create." Anthony pondered this, the idea resonating within him, as they discussed communities of practice and strategic networking. It wasn't just about creating a product; it was about nurturing an ecosystem in which it could thrive.

As Anthony watched Natalie articulate her thoughts—a symphony of insight—he felt as though he was glimpsing a road map that led beyond mere success to significance. He recognized that it was not only about AI; it was about humanity interfacing with it. In the echo of Natalie's words, Anthony saw a future paved by the mentorship before him, a path he was now emboldened to tread.

How does one honor and integrate the guidance from those who have traversed these paths before, while carving one's own legacy in the era of artificial intelligence and machine learning?

Thriving in the AI-Led Professional Landscape

As we venture deeper into the era of artificial intelligence, the sanctity of human connection in career growth remains as vital as ever. There's a Biblical truth in the idea that *"iron sharpens iron"* (Proverbs 27:17), and in modern terms, this wisdom underscores the power of strategic networking and mentorship—especially in the complex world of AI.

The professionals who master the art of fostering relationships with industry forerunners and seeking counsel from experienced mentors are the ones who will carve paths of lasting success and relevance in their fields. Networking and mentorship are not mere buzzwords but timeless principles, now magnified by the urgency of an AI-shaped future.

The fabric of our professional landscapes is undergoing a profound transformation, threaded with the strands of artificial intelligence. With this shift, there's a divine calling to not just adapt passively but to actively seek out connections from those who tread the path before us that illuminate and guide. Whether you're a nascent entrepreneur, a seasoned executive, or a curious professional, building communities of practice related to AI adoption is a step one cannot afford to overlook.

For indeed, as Ecclesiastes 4:9-10 suggests, two are better than one, for they have a good reward for their toil. Mentorship serves as a beacon, guiding professionals through the evolving complexities of AI in their careers. A mentor who has navigated the intricacies of technological disruption can provide tailored advice, critical foresight, and emotional support—a triad of mentorship that enriches our career journey. Navigating the unchartered waters of AI employment, one requires the steady hand of a mentor, a navigator who has charted these waters and can safely lead us through.

In this AI age, knowledge alone is not enough; it's the application of knowledge through strategic insights gained from networking that truly propels professional growth. This implies not merely accumulating contacts but deeply engaging with visionaries and innovators who breathe life into AI concepts and translate them into actionable trajectories. Through investing in such interactions, one can glean a wealth of knowledge and cultivate an agile mindset, essential for ongoing learning and professional versatility.

Behind such strategies lies a well-rounded perspective not confined to technology alone. Like a tapestry woven from varied threads, insights from a cross-section of theology, politics, and economics play integral roles in shaping a holistic approach to career resilience in the AI age. For instance, understanding the political landscape can provide clues to impending regulatory changes affecting AI adoption, while economic trends may reveal future job markets and opportunities for innovation.

As we engage in this meaningful dialogue, remember that the approach is twofold: **to be both teacher and student in the grand classroom of professional development**. Share your journey and lessons with openness, creating a space where wisdom can be exchanged and valued. In this way, the advance of AI becomes not a harbinger of job insecurity but a canvas for evolving one's professional value.

By building connections, professionals can gain valuable insights, expand their knowledge, and position themselves for success in a continuously evolving marketplace. Industry leaders have the expertise and experience to offer guidance and mentorship, providing invaluable support for those seeking to navigate the complexities of AI in their respective fields. Moreover, engaging in communities of practice allows professionals to collaborate with peers, share best practices, and stay abreast of the latest developments in AI technology.

One powerful method for connecting with industry leaders is through networking events, conferences, and seminars focused on AI and technology. Attending these gatherings provides an opportunity to engage in meaningful conversations, make lasting impressions, and build relationships with influential figures in the industry. Similarly, active involvement in online communities, such as AI-focused forums, social media groups, and professional networking platforms, can facilitate connections with like-minded individuals and thought leaders.

Cultivating these relationships is crucial for professionals seeking to future-proof their careers in the AI age.

In addition, engaging in communities of practice related to AI integration offers a wealth of benefits. These communities provide a platform for sharing knowledge, exchanging ideas, and collaborating on innovative projects. Through active participation, professionals can broaden their understanding of AI applications within their specific industry, gain exposure to diverse perspectives, and stay informed about emerging trends and breakthroughs.

Empowerment through Shared Anecdotes

By sharing their own experiences, mentors can motivate and inspire, showing you that the hurdles and setbacks you encounter are often integral to the journey toward success. Their personal anecdotes can serve as a source of motivation, reinforcing the belief that challenges are opportunities for growth.

Building a Supportive Network

Mentor relationships often extend beyond a one-on-one dynamic. They can introduce you to a broader network of professionals and like-minded individuals who share a passion for integrating AI into their careers. This community of peers can offer diverse perspectives, collaboration opportunities, and emotional support as you navigate the intricacies of AI adoption.

Making Informed Decisions

Having a mentor can help you make more informed decisions about the trajectory of your career as it intertwines with AI. As you encounter pivotal points in your professional journey, your mentor's guidance and personalized advice can steer you toward choices that align with your long-term goals and aspirations, even if that means taking on a new opportunity with a new company.

The Bottom Line

In the context of AI employment, mentorship is not just a luxury—it is a strategic imperative. Seek out mentors who have successfully harnessed AI in their professions, and be open to the wealth of knowledge and wisdom they can offer. Embrace the mentor-mentee relationship as a sacred link, one that can guide, inspire, and deeply enrich your career in the AI age.

A.I. Horizons: A Predictive Framework Model for Strategic Networking

Trend Analysis

At the core of the Predictive Framework Model is Trend Analysis, an indispensable tool for professionals who aim to stay ahead of the curve in the AI-dominated employment landscape. By scrutinizing current industry trends and breakthroughs in AI technology, one begins to fathom the trajectory of innovation.

Immersing oneself in literature, be it research papers, industry reports, or news pieces, serves to broaden understanding. Attending conferences and seminars not only yields the latest knowledge but also offers sacred grounds for networking with peers and thought leaders.

Technology Assessment

After crystallizing insights from trend analysis, Technology Assessment beckons, inviting a careful examination of the potential and feasibility of leading-edge AI technologies. It calls for discernment and wisdom, weighing factors like scalability, reliability, cost-effectiveness, against ethical implications, all play a part in determining the right path forward. This step isn't about mere technical evaluation but about envisioning the societal impact and fostering technologies that align with both organizational goals and moral compasses.

Stakeholder Engagement

Stakeholder Engagement, much like the fellowship within a congregation, fosters a diverse and inclusive dialogue surrounding the future of AI. Engaging industry experts, researchers, customers, and employees in profound conversations ensures a holistic view is formed.

Through an array of interviews, surveys, and workshops, organizations can sow the seeds for collaborative innovation and harvest a bounty of perspectives, enriching the strategic approach with manifold insights.

Scenario Planning

Scenario Planning emerges as a generative process, akin to cultivating multiple crops to ensure a bountiful harvest irrespective of seasonal changes. It's an exercise in foresight, creating alternative future landscapes for AI. As one infuses each scenario with different premises, we witness a panorama of possibilities. This step imparts organizations with the agility of David – preparing to face Goliaths of varying uncertainties that tomorrow may bring.

Risk Assessment

With the scenarios unfurled like a map of possible futures, Risk Assessment becomes the discerning eye of the watchtower. Organizations must survey the landscapes for hazards – potential barriers, ethical quandaries, and socio-economic repercussions that lie en route to AI adoption. This vigilant process mirrors the biblical act of girding one's loins, preparing to face challenges with determination and strategy.

Decision Making and Action Planning

Decision Making and Action Planning are the forge where insights from rigorous risk assessment are hammered into solid strategy. Here, the virtue of prudence is paramount – the organization must weigh options and craft action plans to meet the foreseen trends. This could mean investing in human capital, developing alliances, or dedicating resources to innovation. One must act courageously yet thoughtfully to pave the organization's path forward.

Monitoring and Adaptation

The final component, Monitoring and Adaptation, acknowledges that the path of AI is an ever-winding river. Thus, organizations must stay vigilant, updating their strategies always open to wisdom's call. This necessitates a culture of perpetual learning from others, scenario re-evaluation, and dynamic action plan adjustment – it is the dance of adaptation that ensures resilience in an AI-fueled future.

When networking strategically in the AI age, these seven components of the Predictive Framework Model interact in harmony. Together, they create a chorus that sings with anticipation of the future, guiding actions in the present.

By attentively observing trends and assessing technologies, diligently engaging with stakeholders, and prudently planning scenarios, businesses can stand steadfast like a lighthouse amidst the AI tempest. Undertaking sound risk assessments will fuel decisive actions and enduring success requires a vigilant eye for continual adaptation. With this framework as a compass, the modern-day professional can more successfully navigate the AI landscape, ensuring not only prosperity but also purposeful progression in his or her career.

As we tread this path, let us commit to fostering these connections with diligence, respect, and gratitude. Let us seek mentors with sincerity, engage with industry leaders with humility, and participate in communities of practice with an open heart and mind. By doing so, we open ourselves to a world of knowledge, support, and growth, ensuring that we remain at the forefront of our respective fields in the AI-driven era.

Chapter Eight

Empowering Yourself Through AI Mastery

In the bustling clamor of the open-concept office, sunlight played upon the laminated surfaces of desks sprawling across the space, each a hive of screens and papers. David leaned back in his chair, fingers tented, eyes locked onto the computer screen where lines of code cascaded down like runoff from a digital mountain. He felt the weight of the company's latest project on his shoulders, poised on the cutting edge of an AI revolution.

The discourse of algorithms and neural networks had always been David's liturgy, but the sacredness of the pursuit deepened on a personal level. To him, artificial intelligence was a counselor, its wisdom brought forth by those with the vision to harness it. He imagined his forthcoming venture not as an augmentation of self but as a fellowship with machine, a collaboration to innovate and excel beyond the limitations of his sole human capacity.

The soft hum of the fluorescent lighting was occasionally punctuated by the tap-tap-tapping of keyboards that sounded like gentle rain on an office window, a rhythmic reminder of the work at hand. Across from him sat Sarah, her focus unbreakable, coding what would become the neural pathways of a complex machine learning model.

To David, the world's markets were tempests, always churning, yet wisdom whispered that it was not just about surviving the storm but being a beacon within it. He envisioned himself and his colleagues as modern-day Noahs, building arks not of gopher wood but of data, algorithms, and foresight. The flood was technological disruption, and the ark was a mastery of machine learning that kept one's professional contributions essential and irreplaceable.

In a moment of respite from the numerical incantations before him, David's eyes fell upon a team gathering in the corner, whiteboard awash with mind maps and strategies. He pondered the parable of the talents, where to those who make good use of what they have, more will be given to them. The application was clear – AI proficiency was a talent to be developed, a skill to multiply through investment and innovation. Each idea conceived and executed was a testament to human creativity fueled by artificial cognition.

He would soon present his latest project proposal, one that seamlessly blended human intuition with AI insights. It promised a transformation, not only in the products his company offered but in the very fabric of the workplace. Doubt lingered, briefly, on the outskirts of his certainty. Could he truly effectuate such a profound change? Or was he merely a voice in the technological wilderness, preaching an AI renaissance?

As the clock neared the hour of his presentation, David recollected the valor of David against Goliath, an underdog fortified not by size but by faith. Perhaps it was such faith, in oneself and in the transcendent power of artificial intelligence, that would slay the giants of the coming age.

As he gathered his notes and prepared to step into the meeting room, one question lingered - are we ready to seize the slingshot of innovation and become the humble shepherds of a new technological era?

Unleashing the Potential Within: The AI Revolution

As we stand on the cusp of a technological renaissance, it is evident that artificial intelligence (AI) and machine learning have begun to rewrite the rulebook of professional engagement. Professionals across every industry are recognizing the imperatives to adapt and facilitate a symbiosis with these intelligent systems.

For those who take AI seriously, the rise of AI is not a harbinger of obsolescence but an invitation to augment human skill and creativity. To master AI is to understand its language and potential; it's an empowering process that can transfigure one's career trajectory. Consider how the Parable of the Talents (Matthew 25:14–30) teaches us the value of stewardship and the importance of nurturing the gifts given to us. In a modern context, AI literacy is such a talent bestowed upon professionals. Those who proactively adopt AI and invest wisely in understanding and mastering this new technology ("talents") entrusted to them can thrive and reap more abundant rewards, while those who bury it in the ground may find themselves struggling to keep up.

Adorning your skillset with AI proficiency is not merely about gaining a competitive edge; it's about ensuring relevance in a future where collaboration with intelligent machines will soon become the norm. It's about fostering a workplace where innovation thrives. Embracing this technological advance does not erode human value but enhances it, embedding the individual into the very fabric of future progress. Pursuing AI mastery with vigor and determination will pay dividends in securing one's professional future.

Securing your position as an indispensable contributor amidst technological disruption mirrors the Biblical understanding of building one's house on a rock rather than on sand (Matthew 7:24-27). Being proactive in adopting AI equips you to be resilient against the storms of change. This chapter aims to be your guide in this steadfast construction, providing the strategic insights to fortify your career against the shifting sands of job insecurity and creative destruction.

Central to this discourse is **the principle that your education is never finished**. Your pursuit of AI knowledge should be continuous and marked by a hunger for learning and growth. The marketplace abounds with AI and machine learning platforms; engaging with them can transform the abstract into applicable expertise. This practical wisdom aligns with James 1:22, calling us not to deceive ourselves by being hearers who forget but doers who act.

Embarking on this journey will inevitably involve challenges, but remember, resilience is built through overcoming. Let this chapter be your roadmap to AI mastery, with examples of how professionals can leverage technology to catapult their career value. There are compelling reasons to be an early adopter, applying AI to intricate business problems and becoming the node at which human intuition and machine intelligence converge.

Through these insights, prepare to emerge as a beacon of adaptability and excellence, transcending the typical boundaries of your profession. Harnessing the transformative power of AI is not only about securing your career — it's about embracing a philosophy that espouses growth, wisdom, and unyielding perseverance. It's time to embrace the AI revolution with faith and fortitude, secure in the knowledge that your professional calling extends beyond the horizon of automation.

In today's professional landscape, mastering AI and machine learning is no longer a luxury but a necessity for staying ahead of the curve. These technologies have become integral tools for collaboration and innovation in a wide array of professional settings, from healthcare and education to finance and agriculture. By mastering AI, professionals can augment their skill sets and ensure they remain indispensable in the workplace.

Professionals who master AI and machine learning are better equipped to exemplify innovation in their respective fields. By understanding and leveraging these technologies, individuals can forge new solutions and approaches that were previously inconceivable. **AI has the power to enhance productivity, streamline processes, and unlock fresh perspectives, and mastering it enables professionals to harness these capabilities to drive progress.** In essence, AI is a potent force for collaboration, enabling professionals to work in unison with intelligent machines to achieve results that surpass traditional limitations.

Moreover, AI mastery empowers individuals to understand the potential of these technologies, positioning them at the forefront of cutting-edge developments. Rather than being bystanders in the face of technological advancement, individuals in various industries can take charge of their professional evolution by actively participating in the AI revolution. Commanding these tools allows professionals to co-create with AI, influencing and shaping its impact on their respective fields. This proactive collaboration with technology fundamentally transforms the way industries function, providing professionals with the means to spearhead change in their sectors.

Mastering AI and machine learning opens the door to new prospects, offering professionals the potential to enhance their careers and personal growth. By embracing these technologies, individuals can elevate their value in the workplace and secure their positions as indispensable contributors amidst technological disruption. The acquisition of AI proficiency demonstrates a commitment to ongoing learning and adaptation, which are essential qualities in today's rapidly evolving professional landscape.

Elevating Professional Value

Elevating your professional value through AI proficiency is about more than just adding a new line to your resume. It's about demonstrating to employers and colleagues that you are equipped with the knowledge and skills necessary to leverage cutting-edge technology for the benefit of the organization. AI proficiency enables professionals to analyze complex data, automate repetitive tasks, and generate valuable insights, thereby streamlining processes and enhancing productivity in the workplace.

Becoming an Indispensable Contributor

Instead of being threatened by the potential impact of AI on the workforce, professionals can proactively align themselves with this transformative force, further cementing their relevance and value in the workplace. By leveraging AI as a tool for collaboration and innovation, professionals can future-proof their careers and become pivotal assets in navigating the digital landscape.

In the face of relentless technological disruption, adopting a proactive stance is critical. Consider the example of Joseph in Egypt, who interpreted dreams and planned for seven years of plenty followed by seven years of famine. Similarly, by anticipating the impact of AI on industry trends and preparing accordingly, professionals can create their own "seven years of plenty," putting themselves in a position of strength when others might be vulnerable. AI mastery enables professionals to preemptively solve problems and optimize processes, positioning themselves as visionaries and problem-solvers in their field.

Building a Bridge Between AI and Human Expertise

Collaboration between AI and human expertise elevates the potential of what both can achieve. It's about harnessing the calculating precision of AI and the nuanced judgment and critical thinking of human experience. **Mastering AI means being able to integrate it into one's workflow to complement and enhance human decision-making rather than replace it. This human-AI partnership is the cornerstone of future-proofing your career, as you provide the empathy, creativity, and strategic oversight that AI cannot.**

Leverage AI for Streamlined Decision-Making

AI is not just about automation; it's about enhancing human capacity for making informed decisions. The judgment of King Solomon imparts the need for discernment, a quality that AI can augment by providing insights from vast datasets. By mastering AI, one can interpret complex patterns and make recommendations with increased accuracy and speed. This level of decision support is pivotal in high-stakes environments where data-driven insights can be the difference between success and failure.

Turning Disruption Into Opportunity

Professionals who adeptly navigate the tides of AI are like the builder of Noah's ark—turning a potentially dire situation into a vessel for salvation. By redefining their roles and becoming architects of AI systems, they can guide their organizations through choppy waters. **The ability to transition from a user of technology to a creator and innovator within the AI space is a powerful means of ensuring one's relevance in an evolving job market.**

Reflecting Values in the Era of AI

For professionals rooted in faith and values, mastering AI also presents the unique opportunity to reflect these principles in the technological realm. **AI systems are designed based on the values and ethics of their creators. Thus, contributing to AI development empowers professionals to imbue these systems with values such as integrity, fairness, and stewardship.** In doing so, they influence not just the trajectory of their careers, but the ethical landscape of AI itself.

Fostering AI Fluency as a Standard for Excellence

The mastery of AI need not be seen as an isolated skill but as a facet of professional fluency as critical as proficiency in a foreign language or adeptness in negotiation. It forms part of a standard for excellence, enabling professionals to communicate effectively in the language of technology and lead with the confidence of those well-versed in their field. By arming yourself with AI literacy, you reaffirm your commitment to being a proactive, informed, and adaptable leader on your team and in your company and industry.

As you embark on this journey, let your faith in your ability to master AI be unshakable. Draw strength from the realization that the very fabric of your being aligns harmoniously with the principles of adaptation and growth. Embrace this technological shift not as a burden, but as an opportunity to flourish and thrive.

In your pursuit of mastering AI, remember that your efforts are not solely for personal gain but for the greater good of your professional landscape. Share your knowledge and insight generously, for in doing so, you foster a collaborative environment where the collective prosperity is upheld as the people around you become literate in AI.

Chapter Nine

Accelerating Career Trajectories with AI

D awn lingered in silence over the high-rises of the city, a gentle painter brushing a soft orange hue against the calm blue canvas of the sky. Inside an office building glass tower, Aaron sat still, forefinger and thumb pinching the bridge of his nose, eyes shut against the light of a computer screen blanketed in analytics and projections. In the quiet, there was a sanctity, a space for contemplation away from the unstoppable march of progress just beyond the window, where AI was ceaselessly weaving itself into the fabric of careers and industries.

Aaron thought of the parable of the talents, a lesson from a cherished, worn Bible that sat in his apartment. "To one he gave five talents, to another two, to a third one," he recited silently, finding a parallel in his current predicament. His employer had announced a strategic push towards integrating AI, and within that mandate, Aaron saw not a threat but an opportunity—to multiply his talents, to embrace and to learn.

With every tick of the second hand on the analog clock perched on a nearby wall, Aaron's colleagues buzzed with whispers of AI displacing their roles, the undercurrent of anxiety like an unseen riptide. Yet Aaron, feet planted, as though grounding himself in the midst of an unseen storm, felt an unwavering faith in his own capacity to adapt and grow. The efficient pitter-patter of keypresses filled the air as he began drafting a proposal for a new position that would bridge the gap between human insight and AI's analytical prowess: **an AI integration specialist**. "By understanding, shaping, and guiding this new tool," he mused, "I become the steward of my own future."

Throughout the day, his proposal took shape, the tapping of keyboard keys akin to the hammering of a sculptor chiseling away at marble. Aaron crafted each sentence with the precision of a skilled artisan, ensuring that his vision was clear and his intentions were pure. Just as Scripture guided through moral complexities, his plan was to shepherd his colleagues through technological evolution, offering itself as a beacon to those endeavoring in the uncharted waters of AI integration. "There's virtue in this," he thought, weaving ethical threads into his blueprint for change.

His lunchtime was spent in the nearby park, where autumn had laid a carpet of golden leaves underfoot, their rustling whispers a testament to constant change. Aaron observed workers around him, each absorbed in their devices, their lives interweaved with technology in a tacit partnership that echoed his own conviction. Returning to the cool air of the office, a determined energy enveloped him, fuelled by a relentless pursuit of progress tempered by humility.

As the day waned into twilight, Aaron's proposal lay completed, a testament to his resolve and foresight. He thought of the marketplace parables, the wisdom of stewardship and investment, the rewards of diligence. If faith could move mountains, surely it could shape the frontiers of a professional life intertwined with artificial intelligence.

In the enveloping dusk, as the city's lights began their nightly dance, a question lingered in Aaron's steadfast spirit, ready to be shared with those who would listen: If we are given talents in the form of AI, how shall we invest them to yield a harvest worthy of the future?

From Automated Tasks to Automated Promotions

The dawn of AI's influence on career development is upon us, providing an opportunity that is as immense as it is essential for growth. At the heart of this renaissance, professionals stand at the precipice of unprecedented acceleration in their career trajectories. This power surge of potential comes not from overworking or serendipity but from aligning with the ever-advancing current of artificial intelligence.

Embracing AI does not signify a relinquishment of control but the strengthening of one's professional repertoire. It's a strategic embrace of a future where human intelligence and artificial intelligence coalesce to advance individual and organizational success, reflecting the providential design that encourages us to grow in wisdom and stature.

By injecting AI into our skill sets, we become artisans of our own careers, crafting a professional narrative that's not only secure but also dynamic and influential. **The endeavor is not to replace the human experience, but to enhance it, meshing flesh and blood insights with machine-generated efficiency.** This fusion heralds a significant shift in how we define and achieve professional success, informed by data-driven strategies and insights that were once the realm of guesswork or intuition.

Let us not be content to stand still as the tides of technology surge around us, but instead dive headfirst into the AI waters. The six clear steps outlined here crystallize a path to cultivating a powerful AI network that serves as both a bulwark against job insecurity and a catapult into influential roles within any given field.

AI Network Nexus: Your Pathway to Professional Pinnacle

The purpose of this process is to construct a robust framework that interlinks your professional aspirations with AI, auguring a future replete with career advancement. These well-defined, flexible steps guide you to establish and maintain a thriving network in the AI landscape, forging a career arc that bends decidedly upward.

Step 1: Identify Your Allies in the AI Arena

Begin with prayerful discernment, seeking guidance on the AI communities that resonate with your calling. Delve into virtual spheres where AI knowledge converges: online forums, social media groups, and professional organizations devoted to the nurturing of artificial intelligence acumen.

Step 2: Immerse and Impress in Intellectual Interchanges

Join these sanctuaries of intellect and engage sincerely, as you would within your community—sharing, supporting, and learning. Your participation in these communities must be as active and engaging as your professional ambitions.

Step 3: Convene and Connect in Crucibles of Innovation

Allocate patterns of your time rhythmically to both local and global AI gatherings—courses, workshops and seminars, both in-person and virtual, that function as modern-day Agoras where minds meet and wisdom is exchanged. These gatherings are as pivotal to your professional growth as congregational fellowship is to your spiritual growth.

Step 4: Be a Beacon of Knowledge and Generosity

Share your unique insights with your colleagues and your professional network, adding to the AI compendium and in doing so, **establish your reputation as a thought leader**.

Step 5: Forge Alliances in Algorithmic Adventures

With the nurturing of your AI network, seek collaborative projects, merging your human creativity with the cerebral strength of AI. Such partnerships mirror the harmonious existence intended for us—a synergy between our aspirations and the divine forces that guide us.

Step 6: Pursue Perpetual Growth in the Proximity of Pioneers

Artificial intelligence evolves relentlessly; thus, your learning must keep pace with ceaseless fervor. Immerse yourself in the new scriptures of technology—courses, workshops, and literature that continually refresh and expand your own professional gospel according to AI.

By adhering to these steps with faith and diligence, you'll not only insulate your career against the cold winds of automation but also elevate your professional stature to one of irrefutable influence and value. This journey of integration with AI will require commitment, but just as with any worthwhile pursuit, the rewards are proportional to the investment. Let this be your guide, your compass, as you navigate the ever-expanding ocean of artificial intelligence. It promises to be a voyage of both personal revelation and professional revolution.

As professionals, we have the privilege of shaping our careers with intention and strategy. Embracing the adoption of artificial intelligence (AI) as an opportunity rather than a threat can accelerate career growth and enhance our value to the businesses and organizations that employ us. Rather than fearing displacement, we can leverage AI to broaden our skill sets and become indispensable assets in an ever-evolving workforce.

By adopting AI as part of our professional repertoire, we position ourselves for more significant responsibilities, further growth opportunities and promotions, as well as accelerated career trajectories. **As AI continues to shape industries, those who possess the skills to navigate and harness its potential will be in high demand. This paradigm shift presents an opportunity for professionals to position themselves as forward-thinkers, creating the chance to ascend to more influential and impactful roles within their organizations.** Adapting to AI is not only a matter of survival, but an avenue to thrive in a competitive job market.

Embracing AI as part of our skill sets makes us more valuable in business and organizational contexts. **As professionals, our proficiency in AI technologies can increase the efficacy of business**

outcomes, driving innovation and efficiency within the organizations we serve. By intimately understanding and applying AI, we become invaluable contributors to our company's growth and success. Embracing AI strengthens our ability to meet organizational goals and positions us as leaders in fostering a culture of ingenuity and adaptability.

Aligning with Organizational Goals

An essential step in leveraging AI as part of professional skill sets is to align these skills with the strategic goals of the organization. By understanding how AI can contribute to the overall objectives of the business, professionals can tailor their skill sets to directly impact key areas of growth and innovation. This strategic alignment not only demonstrates proactive foresight but also showcases the value of integrating AI within the organizational framework.

Integration with Core Competencies

Integrating AI into professional skill sets should be viewed as complementary to existing core competencies. Rather than replacing human skills, AI enhances and augments them, enabling professionals to perform tasks more efficiently and make data-driven decisions. By merging AI capabilities with their core competencies, professionals can elevate their value within the organization, contributing to improved performance and innovation.

Demonstrating how AI can streamline operations, enhance decision-making, and unveil insights positions a professional not only as a team player but as an instrumental agent of innovation. Every tool in a craftsman's belt adds to the value of their work; similarly, AI in the hands of a savvy professional can lead to fine-tuned processes, insightful analytics, and innovative solutions. Those who adeptly use AI may find themselves positioned for new leadership roles that require overseeing these advanced tools and the teams that leverage them.

By spearheading AI initiatives, professionals can reshape their roles within their companies, ensuring their contributions remain critical linchpins amidst the technological evolution. This strategic positioning can make one indispensable, as they become the key person managing and integrating AI strategies that keep the business competitive.

The Advantage of Ethical AI Use

Finally, a faith-based or principle-based approach to professional development emphasizes the ethical use of tools such as AI. In turning technology towards noble purposes, we not only act responsibly but also set an example in our workplaces. Ethical leadership, especially in the realm of AI, could pave the way to roles that place a higher emphasis on the moral implications of business decisions—roles that are both **essential and revered** in the modern corporate landscape.

In summary, AI integration, when approached as an opportunity for growth and a tool for expanding our God-given capacities, can lead to both professional advancement and greater job security. By recognizing this and moving forward with humility, wisdom, and the right intentions, one can turn the challenge of AI into a stepping stone toward professional fulfillment and secure employment.

Through this integration, we exhibit adaptability, learning agility, and a proactive approach to evolving industry demands, echoing the words of Galatians 6:9, *"Let us not become weary in doing good, for at the proper time we will reap a harvest if we do not give up."* As we persist in integrating AI and enhancing our skills, we position ourselves to reap the rewards of continued career growth and professional development, undeterred by the changing landscapes of industry and technology.

Chapter Ten

Thriving Through Continuous Growth and Adaptation

U nder the hum of neon lights in a bustling co-working space, Jacob sat wedged between the harmony of clacking keyboards and the calming sips of herbal tea. A network of servers hummed like a choir in the corner, a testament to the omnipresence of AI. He perused through lines of code on his screen, each an intricate dance of logic and creativity — a symphony composed by human and artificial intelligence alike.

Jacob's mind danced with prophetic visions — passages from Isaiah 43:19 that spoke of doing a "new thing" springing forth, much like the dawn of AI in the realm of business. The scripture fueled his drive to reshape, to learn the language of machines, to position himself at the intersection of faith and technology. It was a deliberate practice, partaking in continuous learning, where each new programming skill acquired felt like scripture memorized, eternal and powerful.

As Jacob sifted through market trends and economic models, he blended insights like a masterful sermon, intertwining theological reflection with political acumen and economic savvy. It was his ministry to navigate this terrain, armed with the fortitude of faith and the agility mandated by a rapidly evolving industrial landscape.

At times, a mentor's voice would echo through the rhythm of his thoughts; a voice that encouraged not just business acumen but spiritual fortitude. It reminded him of the hidden parables in market fluctuations, and the divine wisdom in understanding the ebb and flow of opportunity. Jacob learned to embrace the transition, to ride the waves of change as one would trust in the steadfastness of God's plan.

When the day waned and the neon lights flickered gently, signaling the time for rest, Jacob packed up, leaving behind the technological temple for the night. He stepped out into the dusky evening, where the world whispered with the rustle of leaves and the distant laughter of children. A question lingered in the air, mingling with the cool breeze — if the essence of thriving in an AI-driven world is continuous growth and adaptation, could it be that God, in His infinite wisdom, has laid out this path of relentless change not as a trial, but as a testament to humanity's potential for boundless evolution?

The Genesis of Reinvention

In a world where change is the only constant, those equipped with resilience and adaptability are the ones who write history. Scripture reminds us that *"to everything there is a season, and a time to every purpose under the heaven"* (Ecclesiastes 3:1). Just as the seasons transition seamlessly, so must we, especially in our professional lives. As we step through the doorway of unprecedented technological evolution, it is crucial that we harness the dexterity to evolve with it. Our capacity for ongoing skill development and adaptation—particularly in the realm of artificial intelligence—is not just beneficial, but essential.

The terrain of the future workplace is being sculptured by AI's relentless chisel. Throughout this book, we've unwrapped layer by layer the manifold intricacies of AI and learned to master the right growth mindset and the necessary strategies not as novelties but as vital instruments of survival. Remember that when Goliath stood formidable, David did not shy away—instead, he chose the right stone. Likewise, we must select the proper skills and strategies to secure our place in the professional battleground.

Cultivating resilience in the face of AI's march requires a foundation of faith in one's own abilities and the knowledge that endurance stems from within. The journey toward mastery and the securing of one's career hinges on an unyielding commitment to growth—a growth that echoes the parable of the talents, where investment and stewardship are rewarded (Matthew 25:14-30).

Similarly, embracing the need for continuous growth underscores our resolve to not only understand but to integrate AI into our professional lives. As craftsmen of our destiny, it is our responsibility to fine-tune our expertise perpetually, much like tending a garden that requires constant care to flourish. We must approach AI with a **growth mindset**, viewing every algorithm, prompt, and tool as an opportunity for expansion and every data set as a blueprint for innovation.

Strategic networking also plays a pivotal role in our quest for future-proof careers. Like a tapestry interwoven with numerous threads, our professional network should be diverse and robust, capable of providing insights and opportunities. As Proverbs 27:17 reflects, *"Iron sharpeneth iron; so a man sharpeneth the countenance of his friend."* In a mutual exchange of knowledge and skills, we position ourselves for collective elevation.

The utilization of AI in our professional endeavors is not mere convenience—it's a strategic imperative. It's about crafting solutions that resonate with the heartbeat of efficiency and effectiveness. **It's about leveraging technology to metamorphosize our roles from being replaceable cogs to indispensable creators.** Strategy, when intertwined with technology, breeds unprecedented possibilities.

This narrative has equipped you with an understanding of AI and machine learning that goes beyond the foundations, transcending into the realm of strategic application. It has illuminated pathways to enhance your skills, amplify your productivity, and transcend the ordinary. Indeed, the power of acquiring and applying this knowledge can safeguard not only your career but also embolden your journey towards professional valor.

Therefore, let us proceed with the confidence that comes with divine wisdom, the courage to master the digital realm, and the conviction to continuously sow seeds of growth. Your career is a garden awaiting your nurture—a place where, with your efforts, diligence, and trust in the Almighty and in yourself, flourishing is not a possibility, but a certainty. Let your work be an act of worship, your growth a testament to divine potential, and your career an odyssey that defies the storms of change.

In an ever-evolving world driven by artificial intelligence, cultivating resilience and adaptability is paramount for professional success. The ability to bounce back from setbacks and pivot in the face of change is a prized skill set in the modern workforce. As AI continues to disrupt traditional job roles and industries, individuals must embrace the necessity of continuous growth and adaptation to thrive in this dynamic landscape. To do so, developing resilience, honing data literacy, adopting a growth mindset, and leveraging strategic networking are crucial components in positioning oneself for success in an AI-driven world.

Adapting to New Work Environments

The integration of AI often leads to the restructuring of work environments and processes. As such, professionals must cultivate the ability to adapt to new work dynamics, technologies, and methodologies. This requires a willingness to embrace change, learn new systems, and adapt to different modes of collaboration. By being open to change and adopting a flexible approach to work, individuals can position themselves as valuable assets within their organizations.

Anticipating Future Skills Demands

A proactive approach to skill development involves anticipating future trends and the evolving demands of the job market. Understanding the skills that will be in high demand due to AI integration allows professionals to strategically invest in their development. **This may involve acquiring expertise in areas such as data literacy, critical thinking, problem-solving, and interdisciplinary collaboration**. By aligning skill development with future demands, individuals can establish themselves as indispensable contributors in their fields.

Mastering Data Literacy

In a world where data is the lingua franca, **data literacy emerges as a non-negotiable truism for sustained success. This means not just the ability to interpret charts and statistics but also the sagacity to ask the right questions and make inferences from AI-generated insights. Armed with data literacy, your dialogue with AI becomes nuanced and insightful, allowing you to direct its focus and distill its outputs into actionable intelligence and informed decision-making.**

Nurturing a Culture of Innovation

The strategic utilization of AI necessitates a cultural paradigm shift—one where innovation is not sporadic but systemic. **Encourage a climate where experimentation is the norm, and potential failure is viewed not as a setback but as a critical step towards discovery.** In this atmosphere, AI becomes a tool for trial, teaching, and transformation. Let AI serve as a catalyst for the continual renewal of your professional practices, ensuring that they remain future-focused and adaptable.

Embracing Ethical Considerations and Accountability

At the heart of the strategic utilization of AI lies a steadfast commitment to ethics and accountability. AI's ethical use must be paramount, as the data it crunches and the decisions it influences bear significant societal impact. Exercise diligence in ensuring your AI applications uphold integrity, contribute positively to society, and reflect the high-

est moral standards. Through ethical AI use, you assert that technological progress and human values can symbiotically thrive.

Conclusion

As we reach the culmination of our journey, it's crucial to remember that our professional evolution is an ongoing process, much like
the cycles of growth we witness in the natural world. Just as a tree
continuously adapts to changing seasons, so too must we adapt to
the ever-evolving landscape of technology and AI. Through ongoing
education, embracing change, and strategically leveraging AI, we can
ensure that our professional journey is not just secure but also incred-
ibly enriching.

With a foundation built on resilience, an insatiable appetite for
learning, and strategic positioning for success, **professionals can truly future-proof their careers**. It is within this continuous growth
and adaptation that we discover the true power of embracing AI to
elevate our professional value and safeguard our livelihoods and those
of our loved ones.

So, as we embark on the next phase of our professional journey,
let us do so with an unwavering commitment to continuous growth
and adaptation, knowing that with each step forward, we are not just
future-proofing our careers, but also enriching the very essence of who
we are.

*Future-Proof: How to Adopt and Master Artificial Intelli-
gence (A.I.) to Secure Your Job and Career* has illuminated the
path to harnessing artificial intelligence and machine learning, not as
adversaries to be feared, but as invaluable allies in our professional
pursuits to transform us into indispensable linchpins for the workplace.

As you, reader, embark upon the next phase of your professional journey, I urge you to seize the reins of your destiny with unyielding conviction. The principles, frameworks and strategies unveiled in this book are not mere abstractions; they are the very essence of your future triumph. However, it is up to you on how you steward what you learn once you finish reading this book.

Remember the Parable of the Talents.

Embrace the **unyielding spirit of determination** and unwavering **faith in your inherent worth**. Stand firm in the knowledge that, armed with the insights garnered from this book, you possess the very mindset and tools to carve an indomitable path in your career, fortified by the timeless wisdom that guides your steps even amidst the most tumultuous of changes.

In the immortal words of Mahatma Gandhi:

*"The future depends on what you do **today**."*

Let this quote serve as the cornerstone of your resolve as you tread the path toward technological mastery and unyielding professional growth. The future is yours to shape, and with the wisdom and enlightenment garnered from these pages, your destiny is nothing short of luminous.

Embrace the future with the fervor of a seeker who has glimpsed the treasures that await, for you are now equipped with the profound insight and unwavering courage to carve a legacy that transcends the bounds of technological evolution. I wish you well. Now go create your future!

About the Author

What is most important to Philip Blackett and what truly forms his identity is his relationship with his Lord and Savior Jesus Christ. Philip's mission for the rest of his life is to Grow God's People, Grow God's Businesses, and Grow God's Kingdom as a good and faithful steward of all God has entrusted him, while having a positive influence on all who he encounters each day as a Kingdom Man.

Professionally speaking, Philip is passionate about helping entrepreneurs and small business owners grow their dream businesses, while utilizing his skillset in sales, marketing and business development. Previously, Philip served as President of Cemetery Services, Inc., a seven-figure business he bought based in the Greater Boston area. It was "his pleasure" to also serve as a Manager for a Chick-Fil-A restaurant.

At FedEx, Philip previously provided support to several senior Marketing executives (including the current CEO) as a Senior Communications Specialist after working on its Corporate Social Responsibility team. Before FedEx, Philip advised investors on Wall Street in New York City as an Equity Research Analyst for Goldman Sachs, where he helped recommend investments in over 100 publicly traded companies across ten industries.

Regarding his education, Philip graduated from the Southern Baptist Theological Seminary with his Masters of Divinity (M.Div) degree with a concentration in Great Commission Studies. He also earned his MBA from Harvard Business School. In college, Philip graduated from the University of North Carolina at Chapel Hill as a Morehead-Cain Scholar, majoring in Political Science and Economics.

Philip is a Life Member of Alpha Phi Alpha Fraternity, Inc. When he is not actively fulfilling his mission, Philip enjoys reading, watching sports, and raising his twin daughters, Sofia and Elizabeth, with his wife Mayra.

Books by Philip

Disagree without Disrespect: How to Respectfully Debate with Those who Think, Believe and Vote Differently than You

Future-Proof: How to Adopt and Master Artificial Intelligence (A.I.) to Secure Your Job and Career

The Unfair Advantage: How Small Business Owners can Use Artificial Intelligence (A.I.) to Boost Sales, Outsmart the Competition and Grow their Dream Businesses without Breaking the Bank

Jesus over Black: How My Faith Transformed Me into a Conservative within the Black Community

Maverick Lineage: What I Learned about Black Conservatism in America

Bridging the GOP Gap: How the Republican Party can Win Over African American Voters with Inclusivity and Trust without Compromising Values

Connect with Philip

f

facebook.com/PhilipBlackettFB

𝕏

twitter.com/PhilipBlackett

in

linkedin.com/in/philipblackett

⃝

instagram.com/philipblackett

▶

youtube.com/@PhilipBlackett

♪

tiktok.com/@pblackett

Facebook:

https://www.facebook.com/PhilipBlackettFB

X (Twitter):

https://twitter.com/PhilipBlackett

LinkedIn:

https://www.linkedin.com/in/philipblackett

Instagram:

https://instagram.com/philipblackett

YouTube:

https://www.youtube.com/@PhilipBlackett

TikTok:

https://www.tiktok.com/@pblackett

Blog:

https://www.PhilipBlackett.com

www.ingramcontent.com/pod-product-compliance
Lightning Source LLC
Chambersburg PA
CBHW020419130626
46549CB00006B/2647